W9-CSP-432

An Album of Fluid Motion

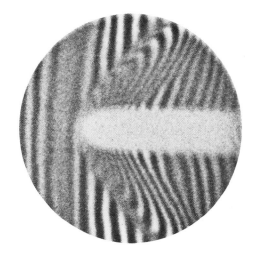

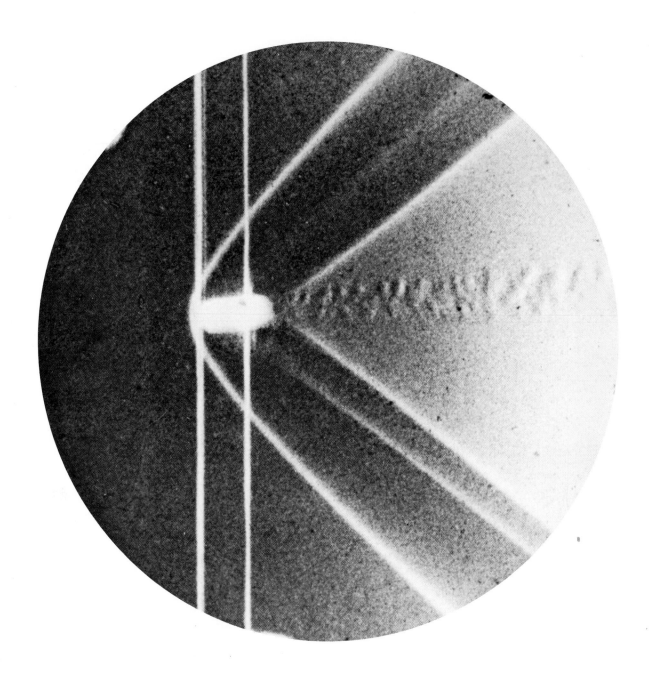

Brass bullet in supersonic flight through air. This photograph, visualized by the schlieren method, was made by Ernst Mach in Prague in the winter of 1888. This print has been enlarged some thirty times, from a negative less than 5 mm in diameter. A year earlier Mach had published the first such photographs ever taken, showing the bow shock wave. Five years later he obtained quantitative measurements of the strength of the shock wave using the device developed by his physician son Ludwig that is now known as the Mach-Zehnder interferometer. *Photograph from the archives of the Ernst-Mach-Institut, Freiburg i. Br., Germany, courtesy of A. Stilp.*

An Album
of Fluid Motion

Assembled by Milton Van Dyke

Department of Mechanical Engineering

Stanford University, Stanford, California

THE PARABOLIC PRESS

Stanford, California

An Album of Fluid Motion

THE PARABOLIC PRESS
Post Office Box 3032
Stanford, California 94305-0030

International Standard Book Numbers:
  Paper: 0-915760-02-9
  Cloth: 0-915760-03-7
Library of Congress
  Catalog Card Number: 81-83088

Printed and bound in the United States of
  America by Braun-Brumfield, Inc.,
  Ann Arbor, Michigan
Set in Goudy Oldstyle by Acme Type,
  San Mateo, California
Designed by Sylvia and Milton Van Dyke

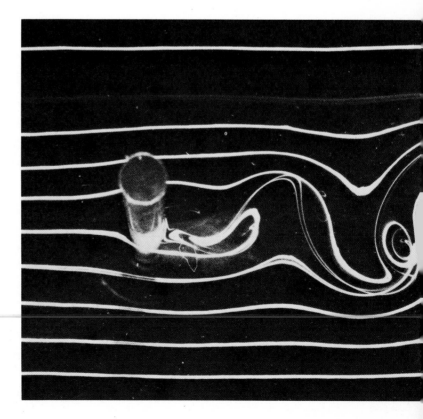

Kármán vortex street behind a circular cylinder.
Smoke filaments in a wind tunnel show the periodic
shedding of vortices. The Reynolds number is about
300, which is near the observed upper limit for stability; and the pattern seems to be disintegrating at
the downstream edge of the photograph. This is an
intermediate stage in the remarkable variety of patterns that succeed one another as the Reynolds number is increased. This stage is preceded by the symmetrical Stokes flow of figure 6 and the symmetrical
standing eddies of figures 40-46, and is followed at
higher Reynolds number by the turbulent wake of
figures 47 and 48. *Photograph by Peter Bradshaw*

Contents

Introduction

We who work in fluid mechanics are fortunate, as are our colleagues in a few other fields such as optics, that our subject is easily visualized. Flow visualization has from early times played an important part in research, always yielding qualitative insight, and recently also quantitative results. Scattered through this century's literature of fluid mechanics is a treasure of beautiful and revealing photographs, which represent a valuable resource for our research and teaching. Most textbooks on fluid mechanics contain at least a few of these photographs.

Since 1958, when I discovered the beautiful *Atlas de Phénomenes Optiques*, I have dreamed of someday assembling a corresponding collection of photographs of flow phenomena. To serve its purpose, however, such a collection needs to be inexpensive enough that it is readily accessible to students. Only recently, when I undertook to republish my own book on perturbation methods, did these twin goals seem within reach. The present collection is not, of course, intended to replace a textbook; but I hope that teachers and students of fluid mechanics will find it a useful supplement.

Colleagues around the world have responded generously to my requests for prints of their best photographs. I asked for one set just as its owner was contemplating discarding it; and a few classics seem unfortunately to have been lost. In particular, I was unable to locate original prints of the beautiful photographs that Prandtl used to illustrate his papers and books. However, I have found modern equivalents for most of those. Only black-and-white photographs appear here. Although I was offered some striking color prints, including them would have—at least in this first edition—defeated my aim of a modest price. Fortunately the great majority of beautiful and informative flow photographs are black-and-white.

As the title of "Album" suggests, this is a rather personal and somewhat haphazard collection. In any case, one cannot arrange the subject of fluid mechanics in any linear fashion. Here the progression is generally from low speeds to high: from creeping flows to shock waves and hypersonic flight. The chapter titles are arbitrary, and many of the photographs could equally well appear in a different chapter.

To minimize confusion, I have flopped negatives where necessary, so that unless otherwise noted a stream flows from left to right and a body is moving from right to left. A sequence of photographs, such as figure 76 on page 43, is arranged to be read from left to right and then from top to bottom, like a comic page. The Reynolds number is based on diameter unless otherwise specified.

Whenever possible I have submitted my legend for revision to the colleague who took the photograph or supplied it, but any errors or shortcomings are my responsibility. The source of each photograph is given at the end of the legend, and if it has been published the reference is cited by surname and year. I have not presumed to convert units, so the legends show a pleasant medley of centimeters and inches.

I anticipate that if this collection is favorably received it will eventually be succeeded by an augmented second edition. I accordingly ask readers to tell me about, and if possible contribute prints of, outstanding photographs old and new that would fill gaps here.

I am grateful to many friends and colleagues who were generous with their encouragement, advice, and help. In addition to the contributors listed in the legends, these include Andreas Acrivos, Holt Ashley, G. K. Batchelor, Amos Clary, Donald Coles, Howard Emmons, J. E. Ffowcs Williams, Sydney Goldstein, Wayland Griffith, Leslie Howarth, R. T. Jones, Frank Kulacki, John Laufer, George K. Lea, James Lighthill, Geoffrey Lilley, E. Rune Lindgren, George H. Lunn, D. W. Moore, M. V. Morkovin, E.-A. Müller, J. Leith Potter, H. Reichenbach, Ann Reynolds, K. G. Roesner, Philip Saffman, Ray Sedney, Helmut Sobieczky, Christopher Tam, Stephen Traugott, Ernest Tuck, K. G. Williams, and Chia-Shun Yih.

MILTON VAN DYKE

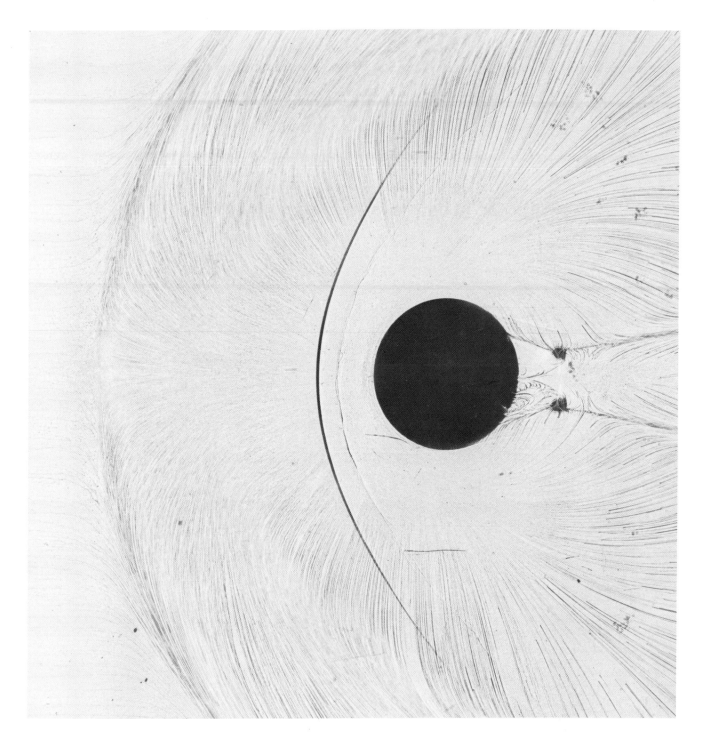

Cylinder extending through a supersonic turbulent boundary layer. This flow combines a number of the phenomena pictured in the subsequent chapters. A circular cylinder of diameter 3.8 cm and height 10 cm is bolted to the window of a supersonic wind tunnel where the turbulent boundary layer is 2.2 cm thick. The free-stream Mach number is 2.50, and the Reynolds number 735,000 based on diameter. A spark light is used to obtain the trace of the bow shock wave as a dark curve typical of shadowgraphs. Upstream, a thin film of lightweight oil on the window marks the line of primary flow separation followed by surface streamlines of the reversed flow in the first of

two horseshoe vortices that wrap around the front of the cylinder. Just downstream of the bow wave, a parallel irregular line indicates the normal portion of an unsteady Mach stem extending into the boundary layer from a triple-shock configuration produced by the separation. Just ahead of the cylinder is the trace of a line of attachment associated with the much smaller second horseshoe vortex. In the near wake, two large dark dots show tornado-like vortices that rise from the surface, bend downstream, and continue as a pair of trailing vortices. *Sedney & Kitchens 1975.*

1. Creeping Flow

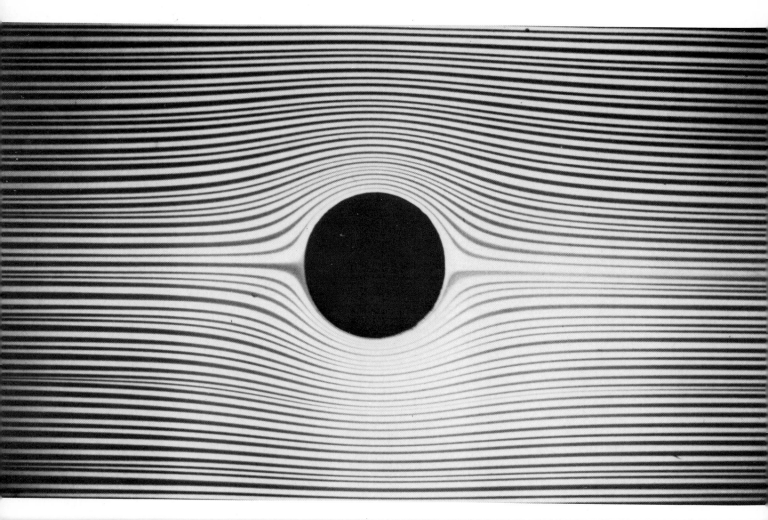

1. Hele-Shaw flow past a circle. Dye shows the streamlines in water flowing at 1 mm per second between glass plates spaced 1 mm apart. It is at first sight paradoxical that the best way of producing the unseparated pattern of plane potential flow past a bluff object, which would be spoiled by separation in a real fluid of even the slightest viscosity, is to go to the opposite extreme of creeping flow in a narrow gap, which is dominated by viscous forces. *Photograph by D. H. Peregrine*

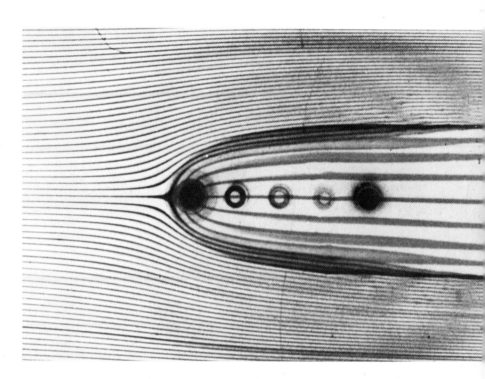

2. Hele-Shaw flow past a Rankine half-body. A viscous fluid is introduced through the orifice at the left into a uniform stream of the same fluid flowing between glass plates spaced 0.5 mm apart. Dye shows both the external and internal streamlines for plane potential flow past a semi-infinite body. The streamlines are slightly blurred because the rate of delivery of fluid to the source was changing as the photograph was made. *Taylor 1972*

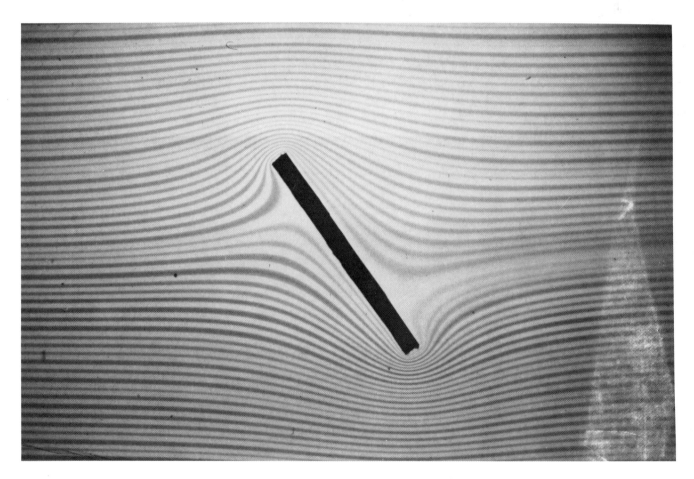

3. Hele-Shaw flow past an inclined plate. The Hele-Shaw analogy cannot represent a flow with circulation. It therefore shows the streamlines of potential flow past an inclined plate with zero lift. Dye flows in water between glass plates spaced 1 mm apart. *Photograph by D. H. Peregrine*

9

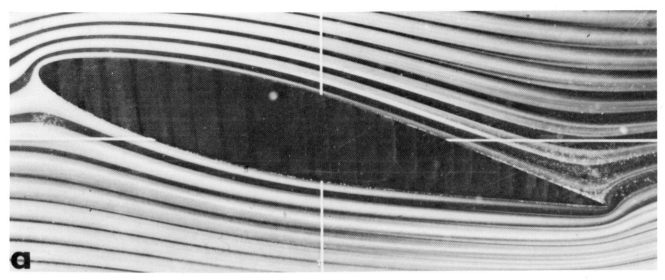

4. Hele-Shaw flow past an inclined airfoil. Dye in oil shows the streamlines of plane potential flow past an NACA 64A015 airfoil at 13° angle of attack. However, because the Hele-Shaw flow cannot show circulation, the Kutta condition is not enforced at the trailing edge. Hence infinite velocities are represented there. The model is between glass plates 1 mm apart. *Werlé 1973. Reproduced, with permission, from the Annual Review of Fluid Mechanics, Volume 5. © 1973 by Annual Reviews Inc.*

5. Hele-Shaw flow past a rectangular block on a plate. The analogy faithfully simulates the unseparated potential flow into the stagnation region of a concave corner, and the infinite velocities over an outside corner. The water takes much longer to travel through the system if it follows a streamline that passes close to a stagnation point. This allows a greater diffusion of dye, which is seen in the slight blurring of streamlines at the lower right-hand corner. *Photograph by D. H. Peregrine*

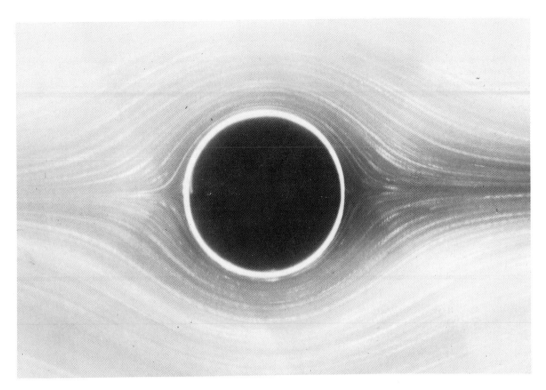

6. Uniform flow past a circular cylinder at $R=0.16$. That the flow is from left to right can scarcely be deduced from the streamline pattern, because in the limit of zero Reynolds number the flow past a solid body is reversible, and hence symmetric about a symmetric shape. It resembles superficially the pattern of potential flow in figure 1, but the disturbances to the uniform stream die off much more slowly. The flow of water is shown by aluminum dust. *Photograph by Sadatoshi Taneda*

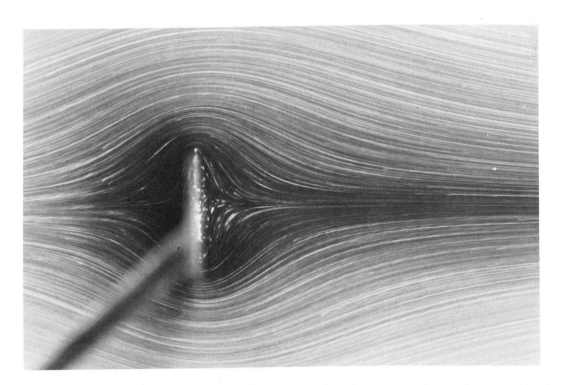

7. Uniform flow normal to a plate at $R=0.334$. The streamline pattern is still almost symmetric fore-and-aft at this higher Reynolds number. It is possible, however, that the flow has separated over the rear. Aluminum dust shows the flow of glycerine. *Taneda 1968*

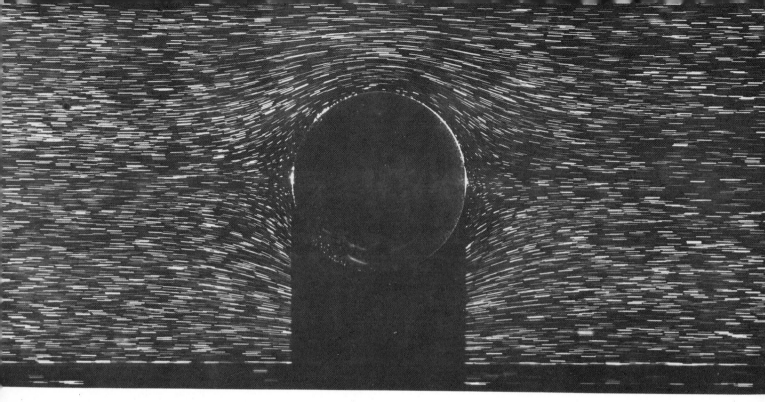

8. Sphere moving through a tube at $R=0.10$, relative motion. A free sphere is falling steadily down the axis of a tube of twice its diameter filled with glycerine. The camera is moved with the speed of the sphere to show the flow relative to it. The photograph has been rotated to show flow from left to right. Tiny magnesium cuttings are illuminated by a thin sheet of light, which casts a shadow of the sphere. *Coutanceau 1968*

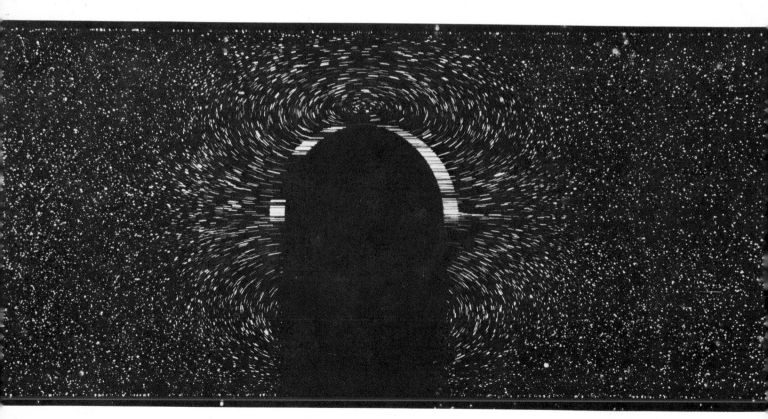

9. Sphere moving through a tube at $R=0.10$, absolute motion. In contrast to the photograph above, here the camera remains fixed with respect to the distant fluid. During the exposure the sphere has moved from left to right less than a tenth of a diameter, to show the absolute motion of the fluid. At this small Reynolds number the flow pattern, shown by magnesium cuttings in oil, looks completely symmetric fore-and-aft. *Coutanceau 1968*

10. Creeping flow in a wedge. The motion is driven by steady clockwise rotation of a circular cylinder whose bottom is seen just below the free surface at the top of the photograph. Visualization is by aluminum dust in water. The Reynolds number is 0.17 based on peripheral speed and wedge height. A 90-minute exposure shows the first two of what are in theory an infinite sequence of successively smaller eddies extending down into the corner. For this wedge, of total angle 28.5°, each eddy is 1000 times weaker than its neighbor above. The third eddy is always so weak that it is not certain that anyone has ever observed it. *Taneda 1979*

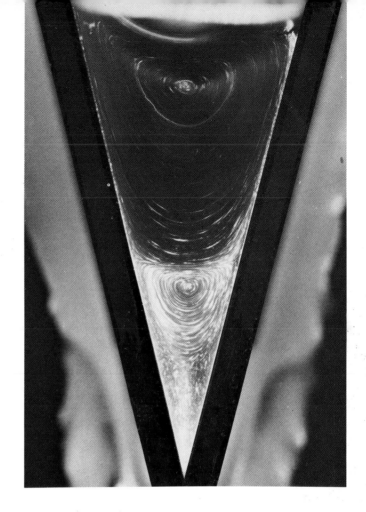

11. Creeping flow past a square block on a plate. The Reynolds number is 0.02 based on the side of the square. In contrast to the unseparated potential streamlines in figure 5, this plane flow separates symmetrically ahead and behind to form large recirculating eddies. Successively smaller and weaker eddies must exist in the corners, as in the wedge above. Visualization is by glass beads in glycerine. *Taneda 1979*

12. Uniform flow past a fence at $R=0.014$. Visualization by aluminum dust in glycerine shows in pure form the separation that occurs on the front and rear of the block in the preceding photograph. *Taneda 1979*

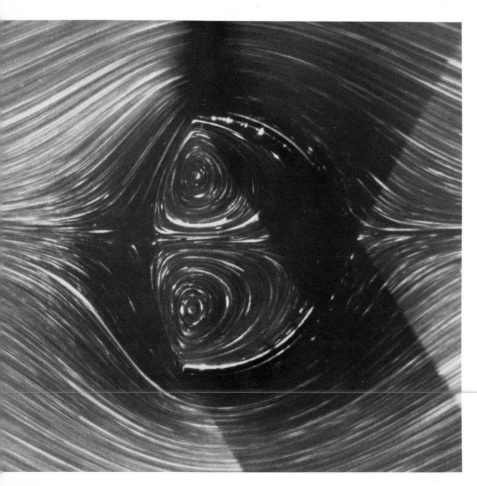

13. Streamlines around a semicircular arc. At this Reynolds number of 0.031 the streamline pattern is not perceptibly altered by reversing the direction of flow. The centers of the pair of eddies in the cavity are separated by 0.52 diameter, in good agreement with a solution of the Stokes approximation. Aluminum powder dispersed in glycerine is illuminated by a slit of light. *Taneda 1979*

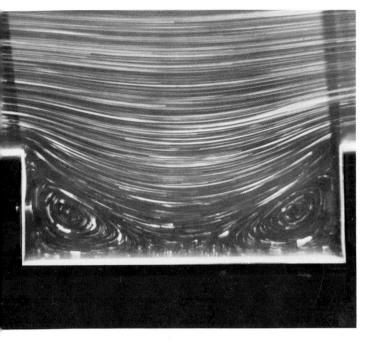

$b/h=3$

$b/h=2$

$b/h=1$

$b/h=0.5$

14. Creeping flow over a rectangular cavity. Streamlines are shown by aluminum dust in glycerine. The Reynolds number is 0.01 based on cavity height. As the breadth of the cavity is reduced, a secondary eddy grows beneath the primary one. If the ratio of breadth to height tended to zero, an unlimited sequence of eddies would form, as in the wedge of figure 10, each weaker than its predecessor by a factor of 365. *Taneda 1979*

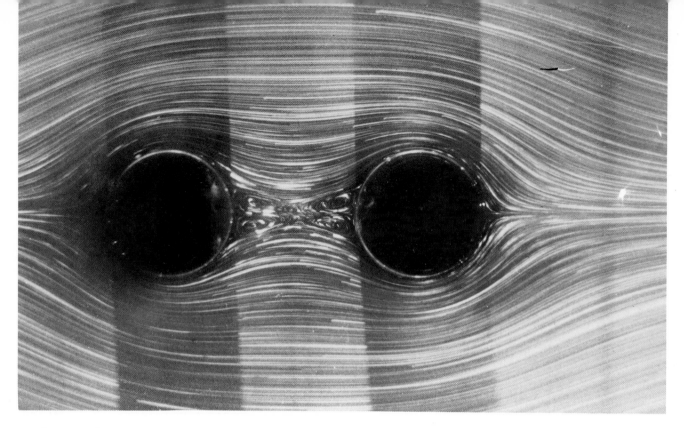

15. Creeping flow past two circles in tandem. The gap is one diameter, and the Reynolds number is 0.01. Streamlines are shown by aluminum dust in glycerine. The inter-action produces separation at any speed, whereas flow past an isolated circular cylinder separates only above a Reynolds number of 5. *Taneda 1979*

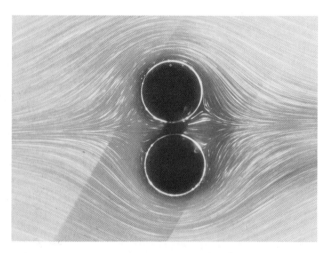

16. Creeping flow past two circles side-by-side. The Reynolds number is 0.011, and the gap between the cylinders is 0.2 of their diameter. Aluminum dust in glycerine shows that there is no apparent separation. *Taneda 1979*

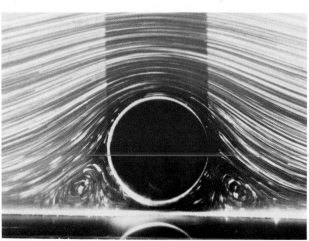

17. Circle in slow linear shear near a plate. The cylinder is 0.1 diameter from the plate, or 0.2 diameter from its hydrodynamic image, which is actually visible as an optical image. The Reynolds number is 0.011 based on the shear rate. Large recirculating eddies form because the glycerine must stick to the plate, in contrast to the photograph above, where it flows along the symmetry plane. *Taneda 1979*

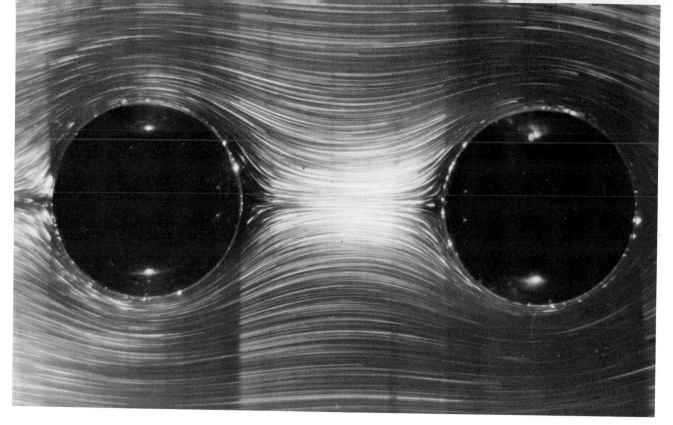

18. Creeping flow past two spheres in tandem. With the same spacing and approximately the same Reynolds number as the circles opposite, spheres show no sign of separation. This is consistent with the fact that separation on an isolated sphere appears only above a Reynolds number of 20, compared with 5 for a circle. Aluminum dust is illuminated in glycerine. *Taneda 1979*

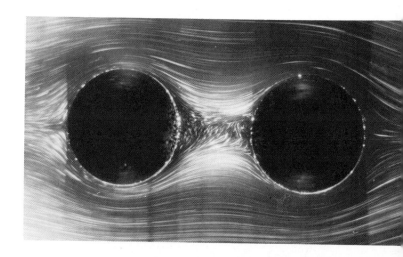

19. Creeping flow past closer spheres. At a spacing of 0.7 diameter, spheres in tandem show separation much like that between the circles spaced one diameter in figure 15. The diameter is 1.6 cm, and the Reynolds number 0.013. *Taneda 1979*

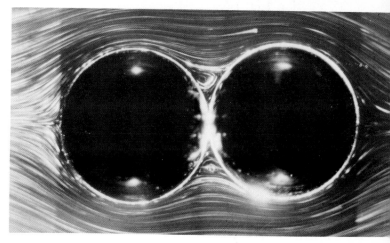

20. Creeping flow past tangent spheres. At the same Reynolds number of 0.013 the two pairs of vortex rings above have now merged into a single pair. Theory predicts, much as for the wedge in figure 10, an infinite sequence of vortex rings nested toward the contact point. *Taneda 1979*

2. Laminar Flow

21. Laminar wake of a slender body of revolution. A sharp-tailed slender body of revolution is supported by fine tungsten wires and accurately aligned with the free stream in a water tunnel. The Reynolds number is 3600 based on maximum diameter. Dye released into the boundary layer shows the core of the wake, which remains laminar to the limit of this photograph. Varicose instability and transition to turbulence occur farther downstream. *Photograph by Francis Hama*

22. Axisymmetric flow past a Rankine ogive. This is the body of revolution that would be produced by a point potential source in a uniform stream— the axisymmetric counterpart of the plane half-body of figure 2. Its shape is so gentle that at zero incidence and a Reynolds number of 6000 based on diameter the flow remains attached and laminar. Streamlines are made visible by tiny air bubbles in water, illuminated by a sheet of light in the mid-plane. *ONERA photograph, Werlé 1962*

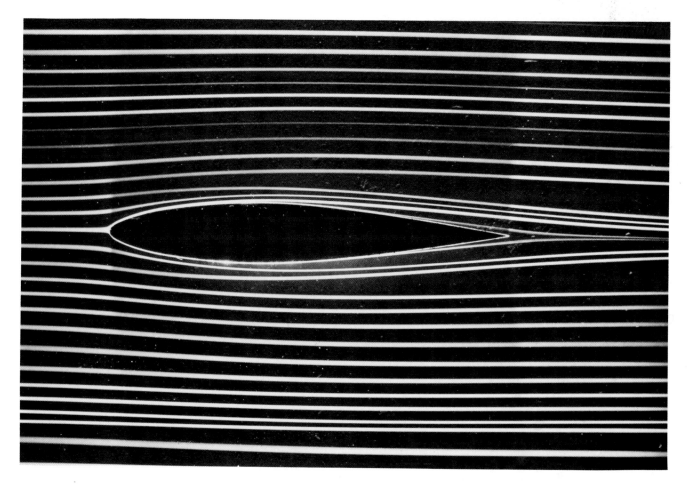

23. Symmetric plane flow past an airfoil. An NACA 64A015 profile is at zero incidence in a water tunnel. The Reynolds number is 7000 based on the chordlength. Streamlines are shown by colored fluid introduced upstream. The flow is evidently laminar and appears to be unseparated, though one might anticipate a small separated region near the trailing edge. *ONERA photograph, Werlé 1974*

24. Circular cylinder at R=1.54. At this Reynolds number the streamline pattern has clearly lost the fore-and-aft symmetry of figure 6. However, the flow has not yet separated at the rear. That begins at about R=5, though the value is not known accurately. Streamlines are made visible by aluminum powder in water. *Photograph by Sadatoshi Taneda*

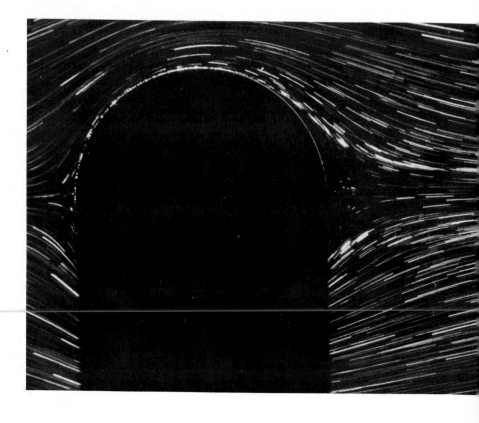

25. Sphere at R=9.8. Here too, with wall effects negligible, the streamline pattern is distinctly asymmetric, in contrast to the creeping flow of figure 8. The fluid is evidently moving very slowly at the rear, making it difficult to estimate the onset of separation. The flow is presumably attached here, because separation is believed to begin above R=20. Streamlines are shown by magnesium cuttings illuminated in water. *Photograph by Madeleine Coutanceau and Michele Payard*

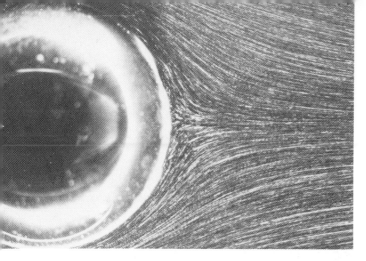

26. Flow behind a sphere at $R=8.15$. A steel ball bearing supported laterally on a fine piano wire is towed through water containing suspended aluminum dust, and illuminated by a sheet of light in the equatorial plane. The flow is clearly not yet separated. *Photograph by Sadatoshi Taneda*

27. Flow behind a sphere at $R=17.9$. As the speed increases it is difficult to discern the onset of separation at the rearmost point. Here the flow must still be attached, because these experiments have indicated that separation behind an isolated sphere begins at about $R=24$. *Taneda 1956b*

28. Sphere moving through a tube at $R=6.9$, absolute motion. The sphere is one-fourth the diameter of the tube. It has moved to the left through one radius. In contrast to the creeping motion of figure 9, there is at this modest Reynolds number the beginning of a wake: the disturbances extend considerably farther behind the sphere than ahead. Magnesium cuttings are illuminated in silicone oil. *Archives de l'Académie des Sciences de Paris. Coutanceau 1972*

21

29. Flat plate at zero incidence. The plate is 2 per cent thick, with beveled edges. At this Reynolds number of 10,000 based on the length of the plate, the uniform stream is only slightly disturbed by the thin laminar boundary layer and subsequent laminar wake. Their thickness is only a few per cent of the plate length, in agreement with the result from Prandtl's theory that the boundary-layer thickness varies as the square root of the Reynolds number. Visualization is by air bubbles in water. *ONERA photograph, Werlé 1974*

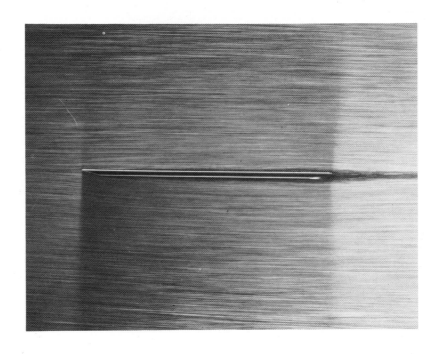

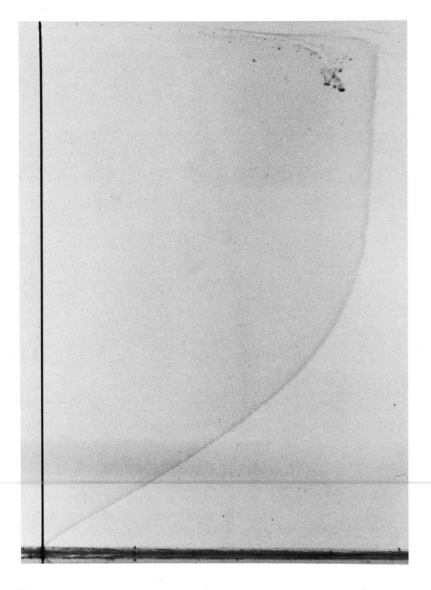

30. Blasius boundary-layer profile on a flat plate. The tangential velocity profile in the laminar boundary layer on a flat plate, discovered by Prandtl and calculated accurately by Blasius, is made visible by tellurium. Water is flowing at 9 cm/s. The Reynolds number is 500 based on distance from the leading edge, and the displacement thickness is about 5 cm. A fine tellurium wire perpendicular to the plate at the left is subjected to an electrical impulse of a few milliseconds duration. A chemical reaction produces a slender colloidal cloud, which drifts with the stream and is photographed a moment later to define the velocity profile. *Photograph by F. X. Wortmann*

31. Secondary streaming induced by an oscillating cylinder. A long circular cylinder is oscillated normal to its axis by a loudspeaker in a mixture of water and glycerine. Suspended glass beads are illuminated in a cross plane by a stroboscope. The amplitude of oscillation is 0.17 of the radius, and the Reynolds number based on frequency and radius is 70. The steady second-order streaming motion is directed toward the body along the axis of oscillation (indicated by arrows) in the inner region, and opposite in the outer region. *Photograph by Masakazu Tatsuno*

3. Separation

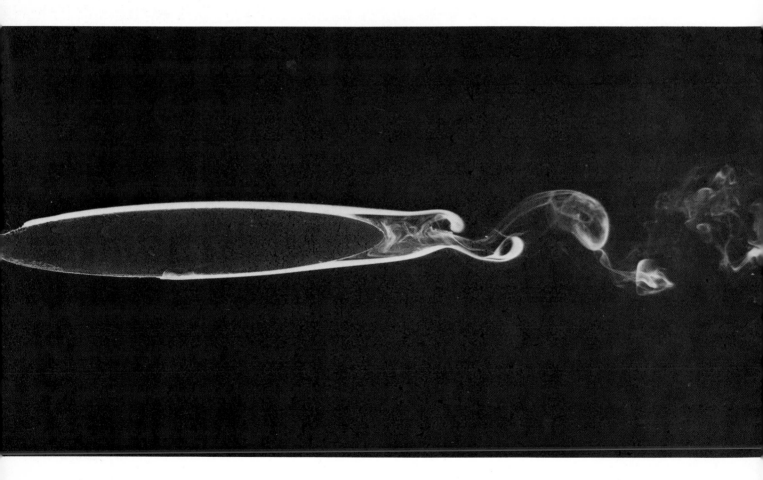

32. Laminar separation on a thin ellipse. A 6:1 elliptic cylinder is held at zero angle of attack in a wind tunnel. The Reynolds number is 4000 based on chord. Drops of ti- tanium tetrachloride on the surface form white smoke, which shows the laminar boundary layer separating at the rear. *Bradshaw 1970*

33. Boundary-layer separation on a body of revolution. This shape is sufficiently blunter than the Rankine ogive of figure 22 that, at the same Reynolds number of 6000 based on diameter and zero incidence, the laminar boundary layer separates. It then quickly becomes turbulent and reattaches to the surface, enclosing a short thin region of recirculating flow. Visualization is by air bubbles in water. *ONERA photograph, Werlé 1962*

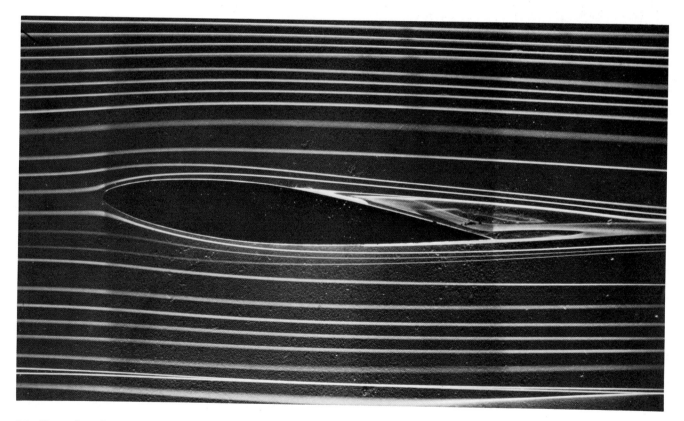

34. Boundary-layer separation on an inclined airfoil. When the NACA 64A015 airfoil of figure 23 is raised to 5° incidence the laminar boundary layer separates from the rear half of the upper surface. The flow remains attached to the lower surface, from which it leaves tangentially at the trailing edge. Streamlines are shown by colored fluid filaments in water. *ONERA photograph, Werlé 1974*

35. Leading-edge separation on a plate with laminar reattachment. A flat plate 2 per cent thick with beveled edges is inclined at 2.5° to the stream. The laminar boundary layer separates at the leading edge over the upper surface. At this Reynolds number of 10,000 based on length it then reattaches while still laminar, enclosing a long leading-edge "bubble" of recirculating fluid. Visualization is by air bubbles in water. *ONERA photograph, Werlé 1974*

36. Leading-edge separation on a plate with turbulent reattachment. The plate is again at an angle of attack of 2.5°, but at a higher Reynolds number of 50,000. The boundary layer now becomes turbulent before reattaching, and as a consequence encloses a short recirculating region. Air bubbles show the flow of water. *ONERA photograph, Werlé 1974*

37. Global separation over an inclined plate. As the angle of attack is increased, the local laminar leading-edge separation shown above spreads rapidly rearward. Here at a Reynolds number of 10,000 and 20° incidence the flow has separated from the entire upper surface. *ONERA photograph, Werlé 1974*

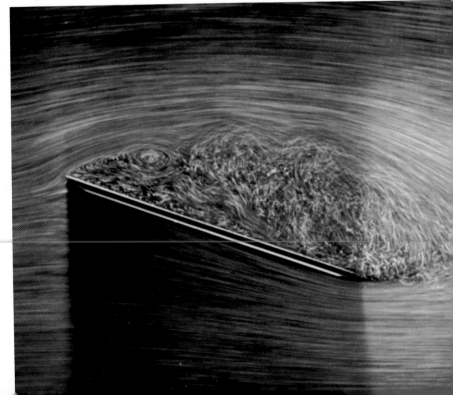

38. Laminar separation from a curved wall. Air bubbles in water show the separation of a laminar boundary layer whose Reynolds number is 20,000 based on distance from the leading edge (not shown). Because it is free of bubbles, the boundary layer appears as a thin dark line at the left. It separates tangentially near the start of the convex surface, remaining laminar for the distance to which the dark line persists, and then becomes unstable and turbulent. *ONERA photograph, Werlé 1974*

39. Turbulent separation over a rectangular block on a plate. The step height is large compared with the thickness of the oncoming laminar boundary layer. The flow is effectively plane, so that the recirculating region ahead of the step is closed, whereas in the corresponding three-dimensional flow of figure 92 it is open and drains around the sides. *ONERA photograph, Werlé 1974*

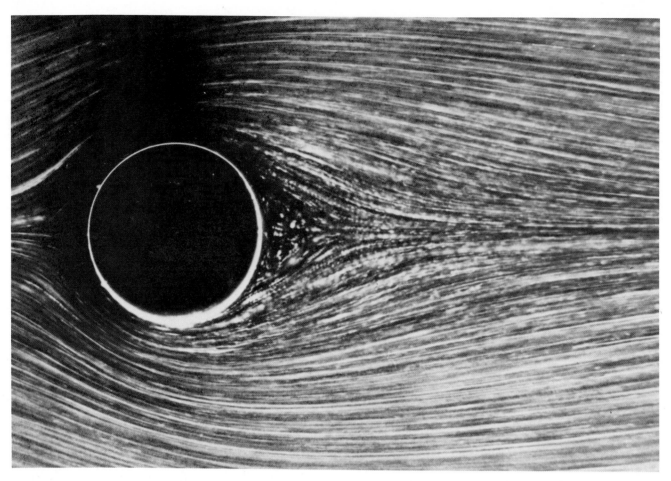

40. Circular cylinder at $R=9.6$. Here, in contrast to figure 24, the flow has clearly separated to form a pair of recirculating eddies. The cylinder is moving through a tank of water containing aluminum powder, and is illuminated by a sheet of light below the free surface. Extrapolation of such experiments to unbounded flow suggests separation at $R=4$ or 5, whereas most numerical computations give $R=5$ to 7. *Photograph by Sadatoshi Taneda*

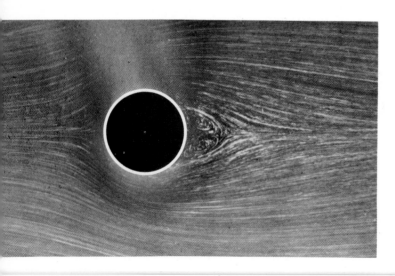

41. Circular cylinder at $R=13.1$. The standing eddies become elongated in the flow direction as the speed increases. Their length is found to increase linearly with Reynolds number until the flow becomes unstable above $R=40$. *Taneda 1956a*

42. Circular cylinder at $R=26$. The downstream distance to the cores of the eddies also increases linearly with Reynolds number. However, the lateral distance between the cores appears to grow more nearly as the square root. *Photograph by Sadatoshi Taneda*

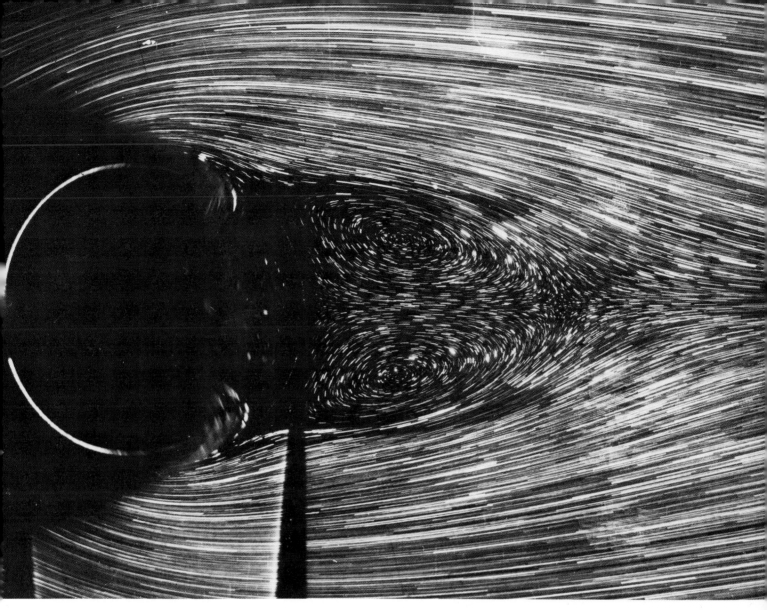

43. Circular cylinder at $R=24.3$. A different view of the flow is obtained by moving a cylinder through oil. Tiny magnesium cuttings are illuminated by a sheet of light from an arc projector. The two dark wedges below the circle are an optical effect. The lengths of the particle trajectories have been measured to find the velocity field to within two per cent. *Coutanceau & Bouard 1977*

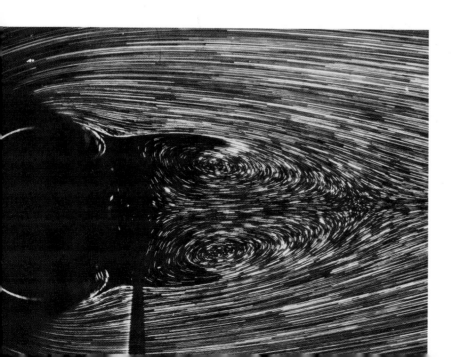

44. Circular cylinder at $R=30.2$. The flow is here still completely steady with the recirculating wake more than one diameter long. The walls of the tank, 8 diameters away, have little effect at these speeds. *Photograph by Madeleine Coutanceau and Roger Bouard*

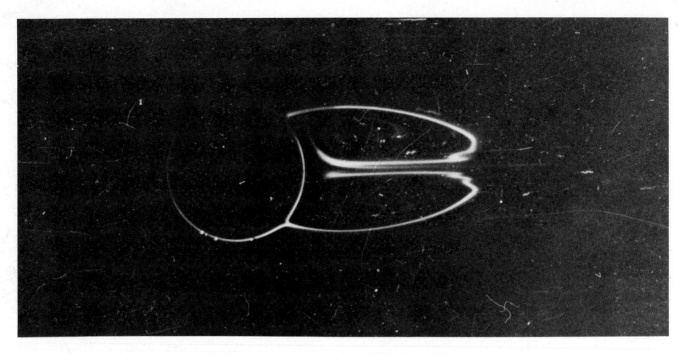

45. Circular cylinder at R=28.4. Here just the boundary of the recirculating region has been made visible by coating the cylinder with condensed milk and setting it in motion through water. *Taneda 1955*

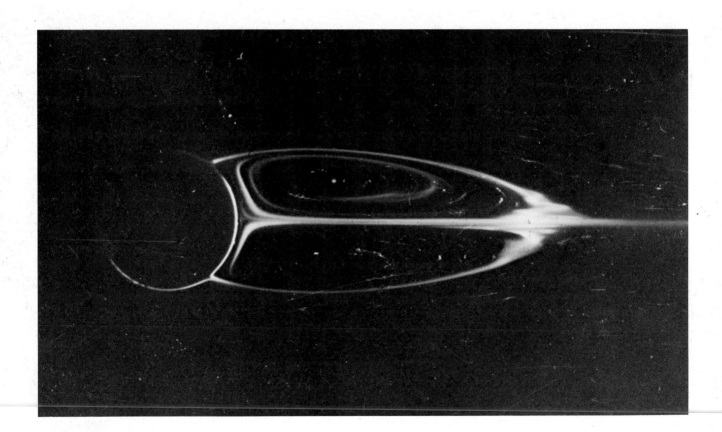

46. Circular cylinder at R=41.0. This is the approximate upper limit for steady flow. Far downstream the wake has already begun to oscillate sinusoidally. Tiny irregular gathers are appearing on the boundary of the recirculating region, but dying out as they reach its downstream end. *Taneda 1955*

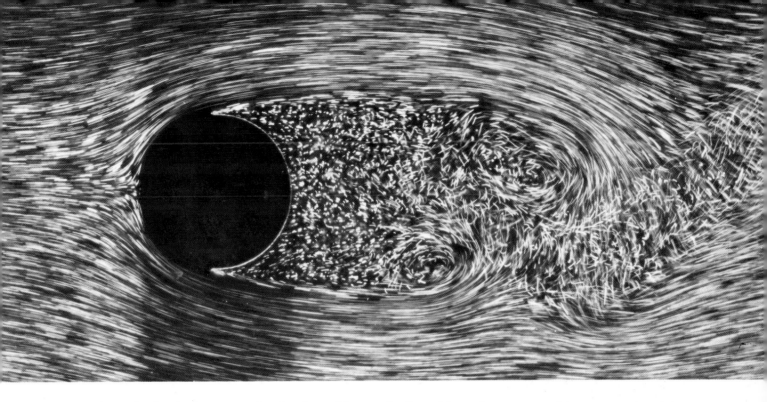

47. Circular cylinder at $R=2000$. At this Reynolds number one may properly speak of a boundary layer. It is laminar over the front, separates, and breaks up into a turbulent wake. The separation points, moving forward as the Reynolds number is increased, have now attained their upstream limit, ahead of maximum thickness. Visualization is by air bubbles in water. *ONERA photograph, Werlé & Gallon 1972*

48. Circular cylinder at $R=10,000$. At five times the speed of the photograph at the top of the page, the flow pattern is scarcely changed. The drag coefficient consequently remains almost constant in the range of Reynolds number spanned by these two photographs. It drops later when, as in figure 57, the boundary layer becomes turbulent at separation. *Photograph by Thomas Corke and Hassan Nagib*

31

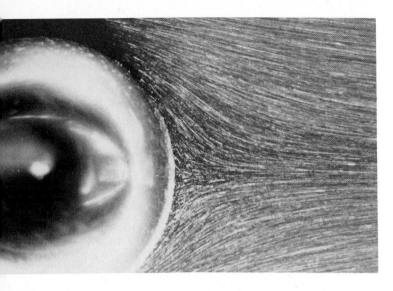

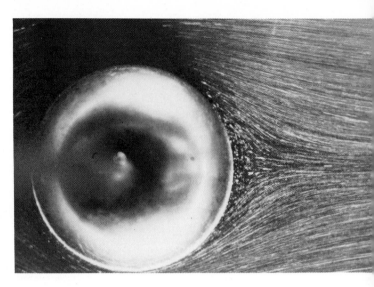

49. Sphere at $R=25.5$. Although it is not obvious, the flow is believed just to have separated at the rear at this Reynolds number, in contrast to the unseparated flow of figure 27. Aluminum dust is illuminated in water. *Taneda 1956b*

50. Sphere at $R=26.8$. At this slightly higher speed the flow has clearly separated over the rear of the sphere, to form a thin standing vortex ring. Aluminum dust is illuminated in water. *Taneda 1956b*

51. Sphere at $R=56.5$. As in figure 8, the sphere is falling steadily down the axis of a tube filled with oil, but here so large that the influence of the walls is negligible. Magne-sium cuttings are illuminated by a sheet of light, which casts the shadow of the sphere. *Archives de l'Académie des Sciences de Paris. Payard & Coutanceau 1974*

52. Sphere at $R=104$. At this Reynolds number the recirculating wake extends a full diameter downstream, but is perfectly steady, as for the circle in figure 44. Visualization is by a thin coating of condensed milk on the sphere, which gradually melts and is carried into the stream of water. *Taneda 1956b*

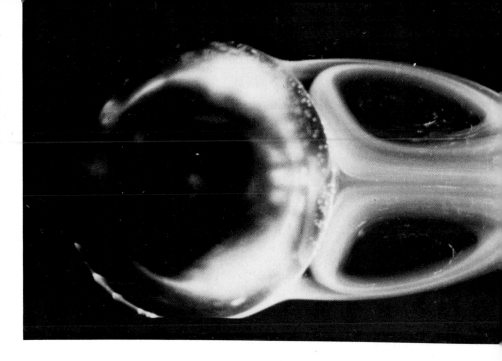

53. Sphere at $R=118$. The wake grows more slowly in axisymmetric than plane flow. These photographs have shown that the length of the recirculating region is proportional to the logarithm of the Reynolds number, whereas it grows linearly with Reynolds number for a cylinder. Aluminum dust shows the flow of water. *Taneda 1956b*

54. Sphere at $R=202$. The rear of the recirculating region behind a sphere begins to oscillate slowly at a Reynolds number of about 130, but the flow is still perfectly laminar at this higher speed. Visualization is by condensed milk in water. *Taneda 1956b*

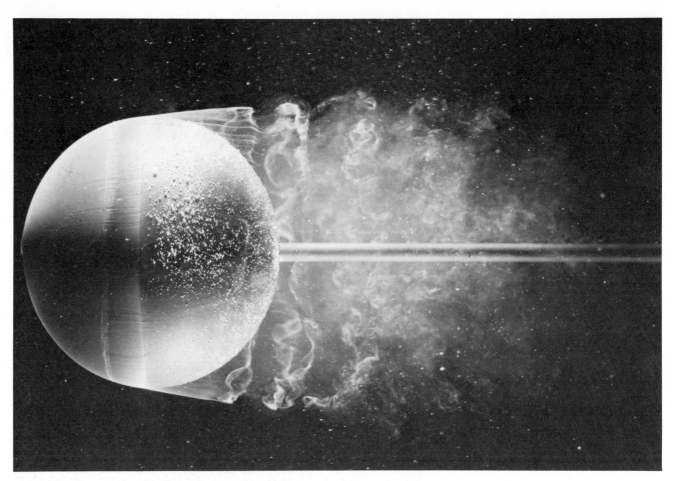

55. Instantaneous flow past a sphere at R=15,000. Dye in water shows a laminar boundary layer separating ahead of the equator and remaining laminar for almost one radius. It then becomes unstable and quickly turns turbulent. *ONERA photograph, Werlé 1980*

56. Mean flow past a sphere at R=15,000. A time exposure of air bubbles in water shows an averaged streamline pattern in the meridian plane for the flow that was photographed instantaneously above. *ONERA photograph by Henri Werlé*

57. Instantaneous flow past a sphere at $R=30,000$ with a trip wire. A classical experiment of Prandtl and Wieselsberger is repeated here, using air bubbles in water. A wire hoop ahead of the equator trips the boundary layer. It becomes turbulent, so that it separates farther rearward than if it were laminar (opposite page). The drag is thereby dramatically reduced, in a way that occurs naturally on a smooth sphere only at a Reynolds number ten times as great. *ONERA photograph, Werlé 1980*

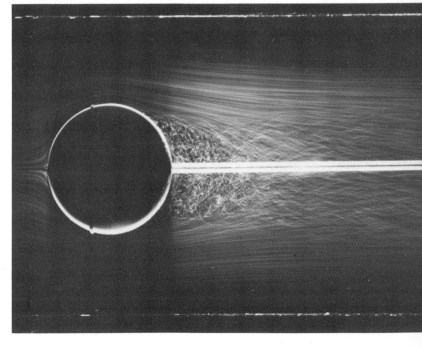

58. Mean flow past a sphere at $R=30,000$ with a trip wire. A time-averaged photograph of the flow above in the meridian plane, visualized by air bubbles in water, shows clearly how the size of the wake is reduced when the boundary layer is turbulent. *ONERA photograph, Werlé 1980*

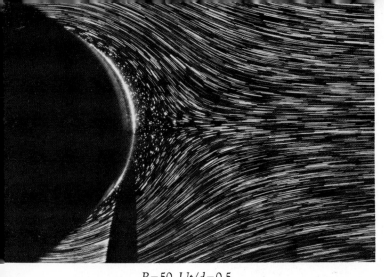

$R=50, Ut/d=0.5$

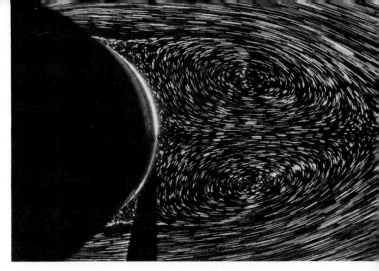

$R=50, Ut/d=2.5$

$R=500, Ut/d=1.0$

$R=500, Ut/d=3.0$

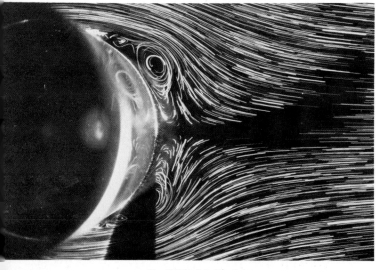

$R=5000, Ut/d=1.0$

$R=5000, Ut/d=3.0$

59. Impulsive start of a circular cylinder. The camera moves with the cylinder, of which only the lighted rear surface is seen. The dark angle below results from a difference in refractive index of the Plexiglas cylinder and the oil through which it is set in motion. The tracer particles are fine magnesium cuttings. *Photograph by Madeleine Coutanceau and Roger Bouard*

36

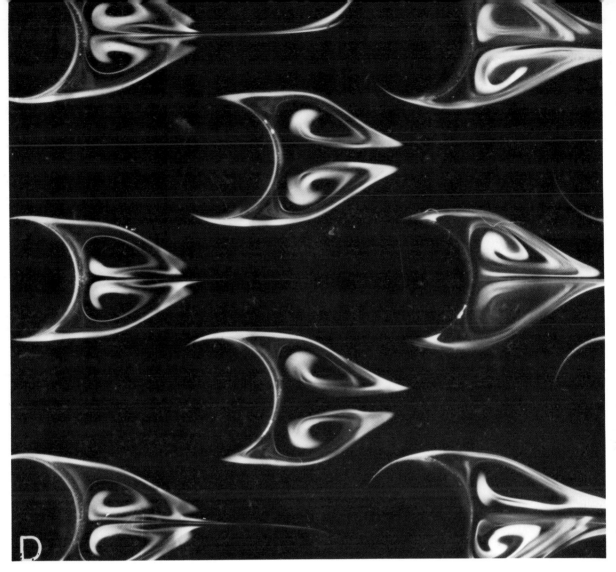

60. Impulsive start of a staggered array of circular cylinders. The Reynolds number is 3000 based on the diameter. The array has moved about ten diameters. *ONERA photograph, Werlé & Gallon 1973*

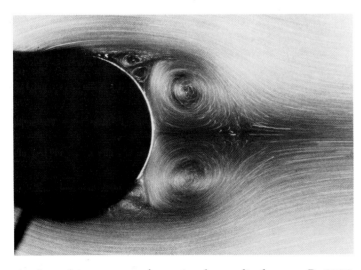

61. Impulsive start of a circular cylinder at $R=1700$, $Ut/d=1.92$. Aluminum dust in water gives a different view of the motion on the opposite page. It clearly shows the pair of small secondary vortices upstream of each main one. *Honji & Taneda 1969*

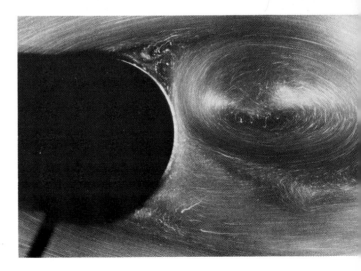

62. Impulsive start of a circular cylinder at $R=1700$, $Ut/d=4.05$. At this later stage the wake has lost its original symmetry, and is beginning to shed vortices into the stream. *Honji & Taneda 1969*

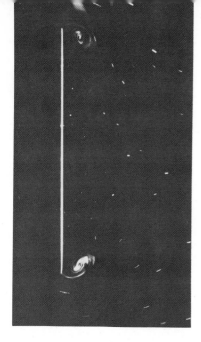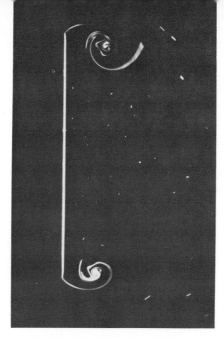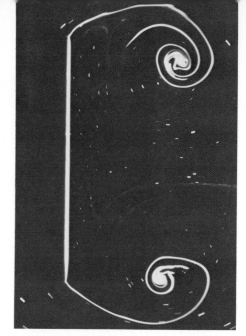

63. Impulsive motion of a flat plate normal to itself. The Reynolds number is 88 based on breadth. White dye generated on the plate by electrolysis of water shows a spiral vortex sheet shed from each edge. The plate has moved 0.079, 0.26, and 0.93 breadths. *Taneda & Honji 1971*

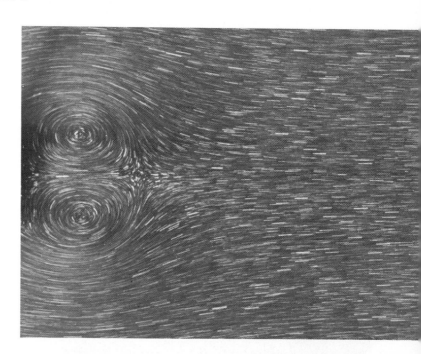

64. Impulsive motion of a flat plate. A quite different view of the phenomenon shown above is given by aluminum dust suspended in the water. It shows a symmetrical pair of standing eddies when the plate has moved 1.45 breadths at a Reynolds number of 126. *Taneda & Honji 1971*

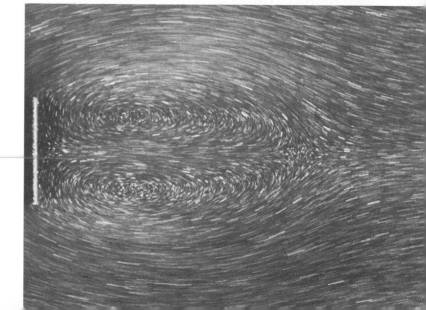

65. Impulsive motion of a flat plate. The plate has now moved from rest 5.02 breadths. The flow pattern is still symmetric, though it will later break down into an oscillating vortex street. The length of the recirculating region is observed to grow as the $\frac{2}{3}$-power of time. *Taneda & Honji 1971*

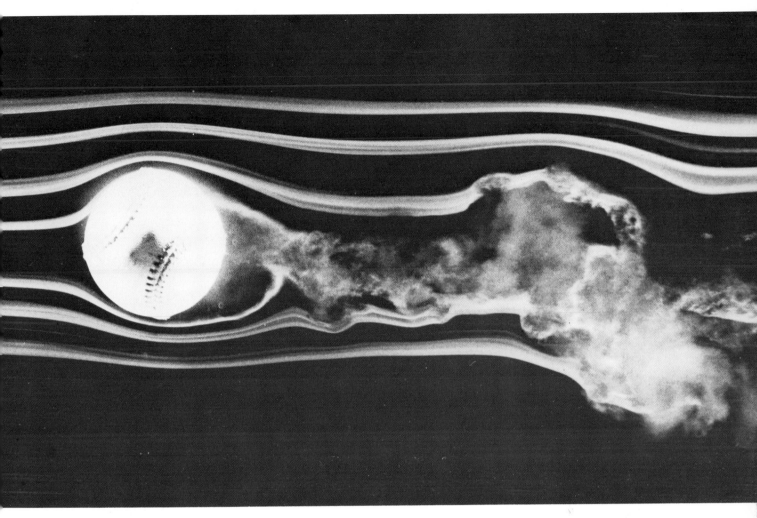

66. Spinning baseball. The late F. N. M. Brown devoted many years to developing and using smoke visualization in wind tunnels at the University of Notre Dame. Here the flow speed is about 77 ft/sec and the ball is rotated at 630 rpm. This unpublished photograph is similar to several in Brown 1971. *Photograph courtesy of T. J. Mueller*

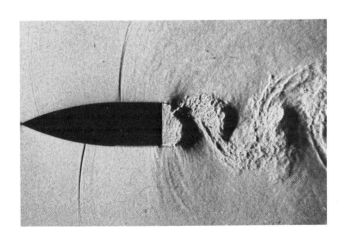

67. Oscillating wake of a blunt-based airfoil. At 0.6 Mach number and Reynolds number 220,000 a high-speed schlieren motion picture shows waves moving upstream alternately over each surface from a periodically oscillating wake. The separation is laminar at the base. *Dyment, Flodrops & Gryson 1982*

68. Mean flow over a blunt-based airfoil. A 1/400-second exposure averages the flow at the left over a dozen cycles to give a completely different impression of the motion. *Dyment & Gryson 1978*

69. Laminar flow up a step. The obstacle spans a 1-mm gap between glass plates, as in the Hele-Shaw photographs of figures 1-5, but water gives a Reynolds number of 1000. The separation pattern is closer to that of figure 11 than of figure 5 or 39. Streamlines are shown by colored fluid. *ONERA photograph, Werlé 1960b*

70. Axisymmetric flow down a step. At a Reynolds number of 10,000 based on body length, the boundary layer is laminar as it separates. The flow is shown by air bubbles illuminated in water. It remains steady and laminar behind the step, becoming unsteady and turbulent farther downstream. *ONERA photograph, Werlé 1974*

71. Wake of a flat-based body of revolution. The laminar boundary layer separates from the base, enveloping a laminar "dead-water" region. Farther downstream it undergoes transition to a turbulent wake. *ONERA photograph, Werlé 1974*

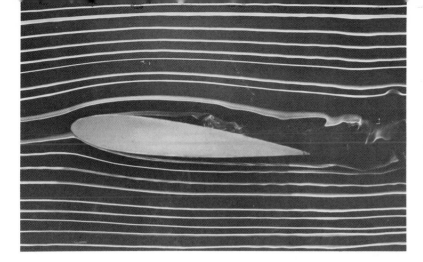

72. Symmetrical airfoil at angle of attack. Smoke in a wind tunnel shows separation over the upper surface of a profile that is 15 per cent thick at 6° incidence and a Reynolds number of 20,000. *Photograph by Peter Bradshaw*

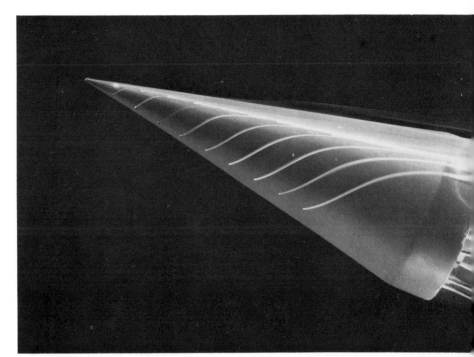

73. Cone at angle of attack. A circular cone of 12° semi-vertex angle is inclined at 16° in water. The Reynolds number is 15,000 based on length. The flow is seen to be sensibly conical. Surface streamlines, shown by colored fluid, are tangential to a thin layer of laminar separation over the leeward face. *ONERA photograph, Werlé 1962*

74. Rankine ogive at angle of attack. The axisymmetric body of figure 22 is here inclined at 30° to a stream of water. Surface streamlines are tangential to the outer edge of a thin layer of laminar separation and reattachment that forms a sort of horseshoe vortex over the leeward surface. *ONERA photograph, Werlé 1962*

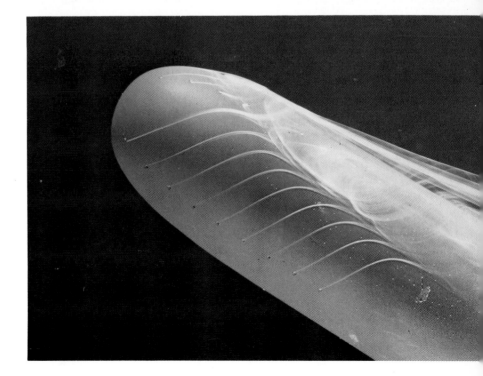

4. Vortices

75. Vortices behind a rotating propeller. A striking pattern of helical tip and root vortices is revealed by smoke in the Notre Dame wind tunnel. The stream flows at 48 ft/s while the propeller rotates at 4080 rpm. *Brown 1971, courtesy of T. J. Mueller*

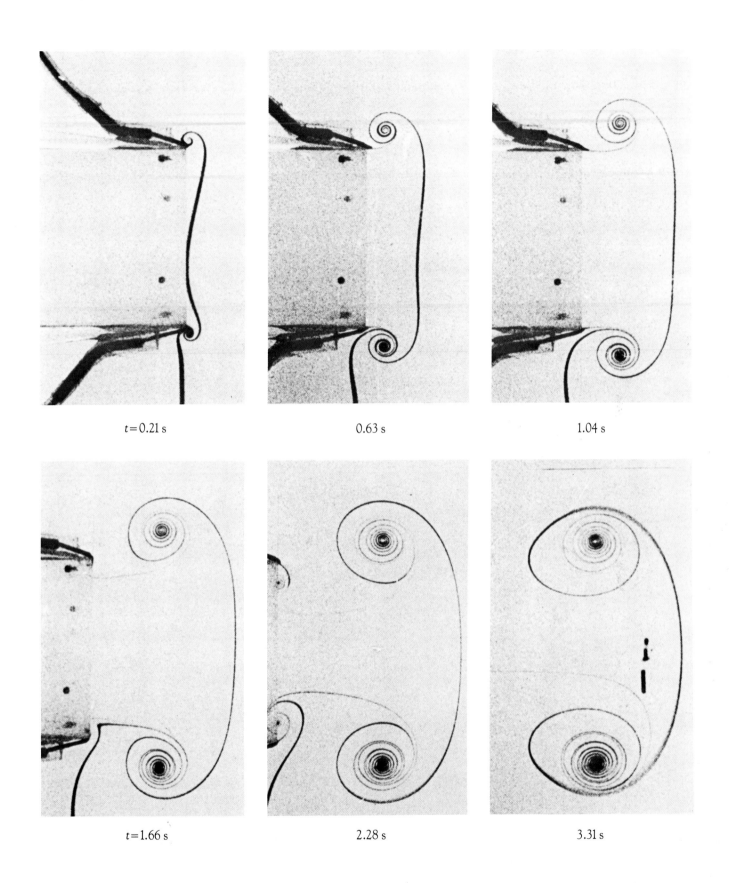

$t=0.21$ s   0.63 s   1.04 s

$t=1.66$ s   2.28 s   3.31 s

76. Formation of a vortex ring from a nozzle. Water is ejected from a sharp-edged circular nozzle of 5-cm diameter into a tank of water by a piston that moves at a constant speed of 4.6 cm/s after accelerating for 0.3 s. The rolling up of the vortex sheet that separates from the edge is shown by dye injected there. The piston stops at 1.6 s, and the vortex ring then induces a secondary vortex of opposite circulation. A different view of this process is shown in figure 112. *Didden 1979*

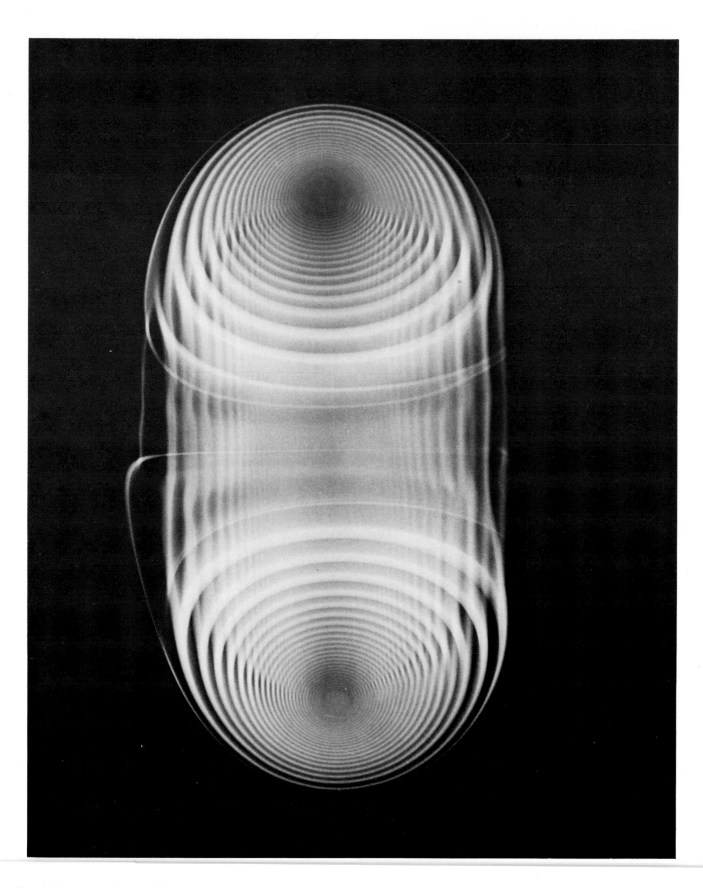

77. Structure of a smoke ring. Ejection of smoke from the end of a tube into air gives a different view of the process shown on the preceding page. It is clear that the result is not a true ring, with a closed surface, but a tightly wound toroidal spiral. The Reynolds number is approximately 10,000. Figure 112 shows the formation of a similar ring in water. *Magarvey & MacLatchy 1964*

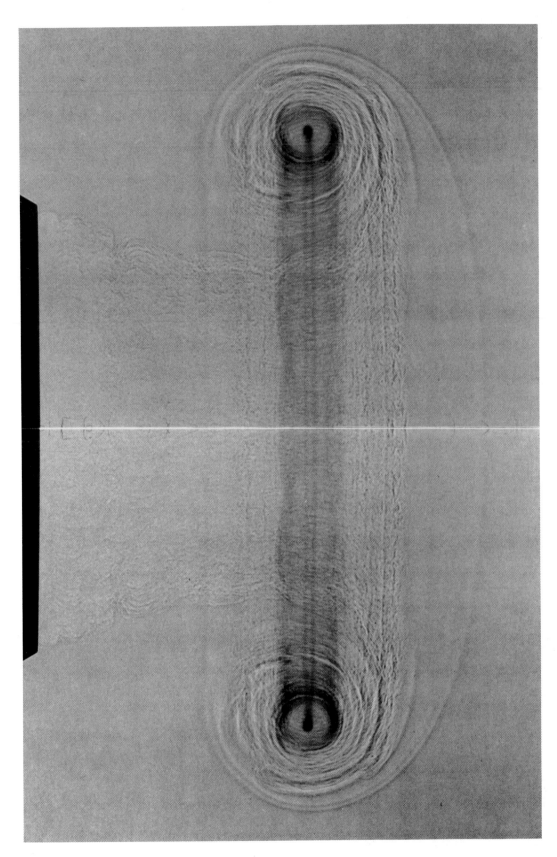

78. Core of a vortex ring. The ring is produced by diffraction of a pressure pulse out of a tube in air. The waves have subsided, leaving the vortex ring moving away from the mouth of the tube. The dark portions of the core are visible in this spark shadowgraph because of the natur- ally occurring reduced density in the rotational flow. In ad- dition, the tube was precooled to aid in the visualization of the outer portions of the ring. Only the upper half was photographed, and then reflected in the axis to give the effect of a complete ring. *Photograph by Bradford Sturtevant*

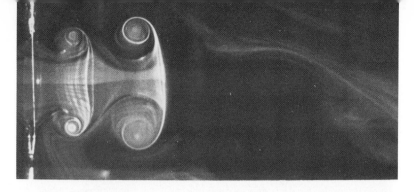

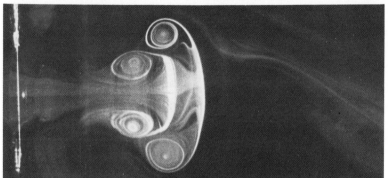

79. Leapfrogging of two vortex rings. Two successive puffs of air are ejected from an orifice of 8-cm diameter by a piston that is driven by the impacts of two pendulums. The flow is made visible by a smoke wire stretched across the orifice, at the left of the photographs. At this Reynolds number of about 1600 based on orifice diameter, the second ring travels faster in the induced field of the first, and has slipped through it in the third photograph. Then the process is repeated, the first ring slipping through the second in the last photograph. *Yamada & Matsui 1978*

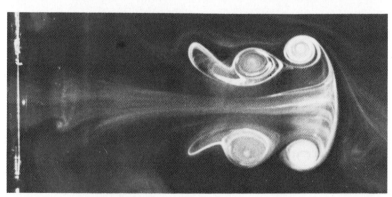

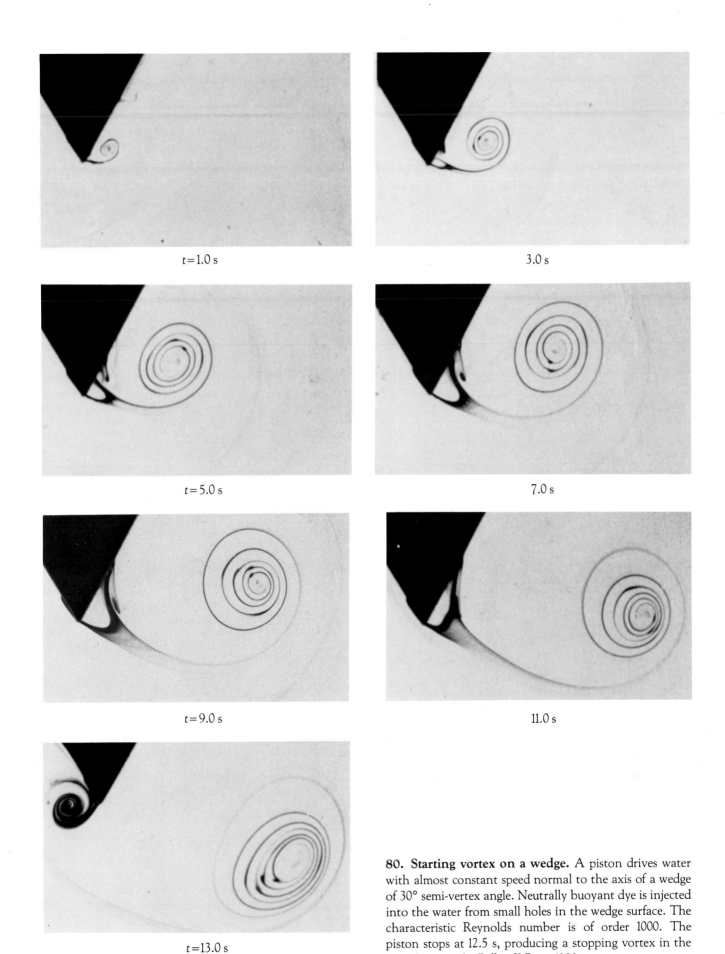

$t=1.0$ s

3.0 s

$t=5.0$ s

7.0 s

$t=9.0$ s

11.0 s

$t=13.0$ s

80. Starting vortex on a wedge. A piston drives water with almost constant speed normal to the axis of a wedge of 30° semi-vertex angle. Neutrally buoyant dye is injected into the water from small holes in the wedge surface. The characteristic Reynolds number is of order 1000. The piston stops at 12.5 s, producing a stopping vortex in the last photograph. *Pullin & Perry 1980*

$t=1.05$ ms, $v=5.5$ ft/s

$t=2.14$ ms, $v=11.1$ ft/s

$t=3.22$ ms, $v=16.9$ ft/s

$t=4.30$ ms, $v=21.0$ ft/s

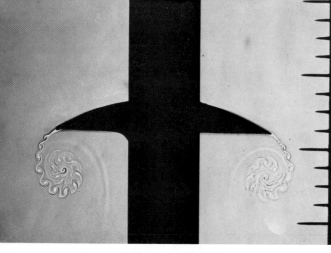

$t=6.53$ ms, $v=24.0$ ft/s

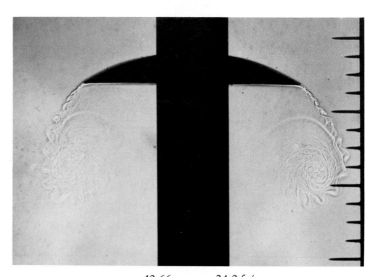

$t=10.66$ ms, $v=24.0$ ft/s

81. Growth of vortices on an accelerated plate. Spark shadowgraphs show the history of a 3-inch-square plate in air, accelerated from rest to 24 ft/s. The sharp edge of the plate is initially opposite the first of a series of pins spaced ¼ inch apart. The motion is actually vertical, and the flow is visualized by painting a narrow band of benzene across the center of the balsa-wood plate, so that when the plate accelerates benzene vapor is drawn into the vortex sheet. The difference in density between the vapor and the air makes the paths of their boundaries visible. Care was taken to ensure that the undulations observed in the vortex sheet were not caused by vibrations of the model. *Pierce 1961*

48

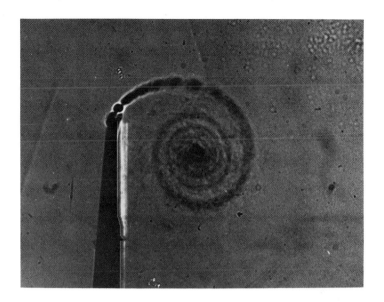

82. Vortex from a wedge in a shock tube. This schlieren photograph shows the vortex that spirals from the tip of a thin wedge after the air is set in motion normal to it by the passage of a weak plane shock wave, which is out of sight to the right. Other photographs show that the flow pattern is "conical" or "pseudo-stationary," remaining always similar to itself but growing in size in proportion to the time. *Photograph by Walker Bleakney*

83. Density in a vortex from a wedge. A quite different view of the phenomenon above is given by this infinite-fringe interferogram, which shows lines of constant density. A striking feature is the almost perfectly circular density distribution about the center of the vortex, extending nearly to the wedge. *Photograph by Walker Bleakney*

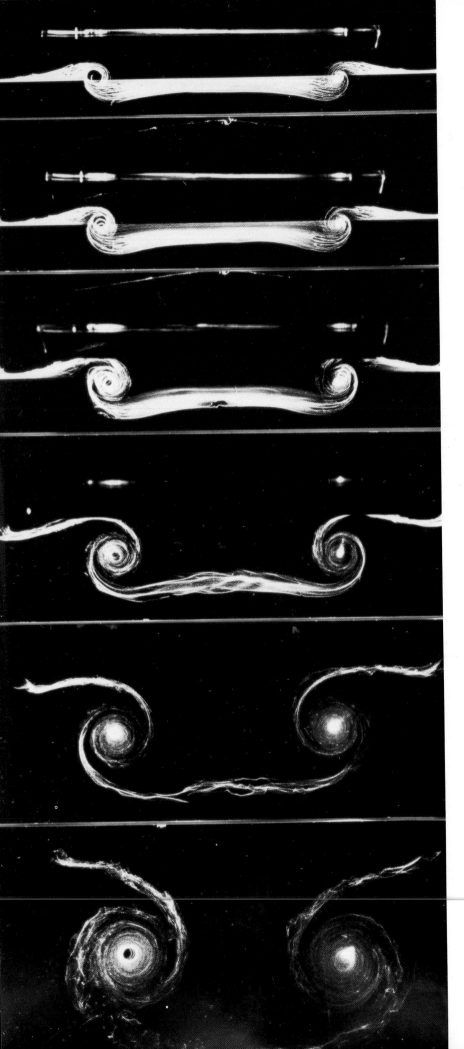

84. Cross-sections of the vortex sheet behind a rectangular wing. The wake rolling up behind a rectangular wing at 9° angle of attack is seen at distances behind the trailing edge of 1.0, 1.6, 2.9, 5.5, 11.2, and 21 chord lengths. The wing has a Clark Y section 10-percent thick, a chord of 0.125 m, and a span of 0.3 m, with tips cut off square. It is supported by fine wires and towed through water. The wake is visualized by hydrogen bubbles emitted from a 30 μm wire located just behind the trailing edge and illuminated by xenon lamps. The vortices separate from the wing surface just behind mid-chord. The Reynolds number is 100,000 based on chord. The vortex sheet is initially turbulent, but is relaminarized farther downstream. *Photograph by H. Bippes*

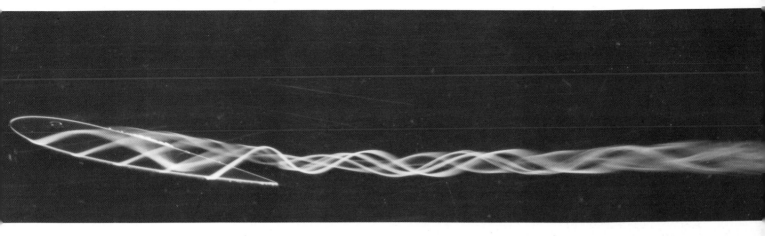

85. Trailing vortex from the tip of a rectangular wing.
At 12.5° angle of attack the vortex is seen to separate well
ahead of the trailing edge. The wing has an NACA 0012
profile and aspect ratio 4. At this Reynolds number of
10,000 the wake is laminar, in contrast to the opposite
page. Visualization is by colored fluid in water. *ONERA
photograph, Werlé 1974*

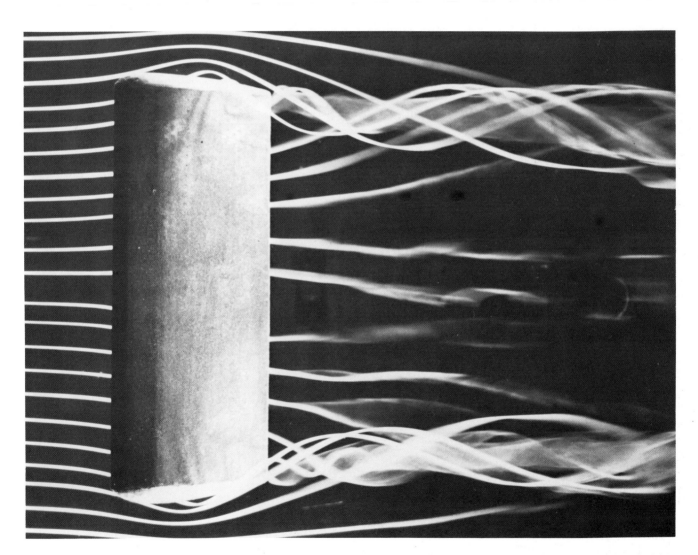

86. Trailing vortices from a rectangular wing. Suction
is applied so that at 24° angle of attack the flow remains attached over the entire wing surface, in contrast to the preceding photograph. The centers of the vortex cores therefore spring from the trailing edge at the tips. The model is
made of perforated metal covered with blotting paper, and
tested in a smoke tunnel at Reynolds number 100,000.
Head 1982

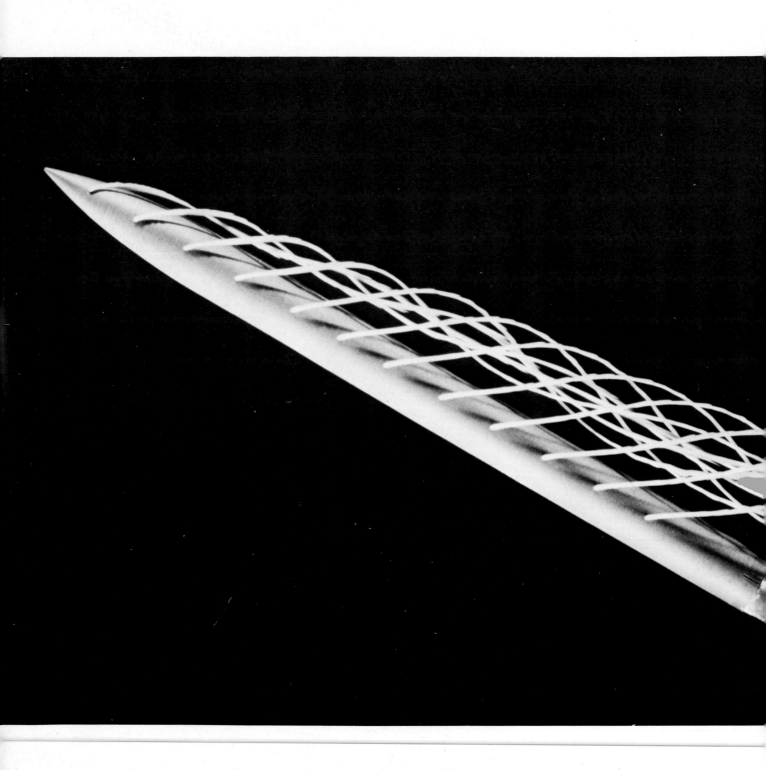

87. Attached vortex pair behind an inclined slender body. A long ogive-cylinder is inclined at 30° to water flowing at 4 cm/s. At this angle of attack a symmetric pair of vortices forms on the lee side of the body. Colored fluid emitted under slight pressure from 0.3-mm holes spirals around the core of the nearer vortex. The Reynolds number is 400 based on the diameter of 1 cm. *Fiechter 1969*

88. Vortices shedding behind a body in subsonic flow. A long cone-cylinder is at 30° angle of attack in a wind tunnel at 0.4 Mach number. The Reynolds number is 80,000 based on the diameter. A symmetric pair of vortices forms in the lee of the forward part, as on the opposite page. Farther downstream, vortices separate alternately, behaving like a Kármán vortex street (figures 94-98) in planes normal to the axis of the body. *Photograph by K. D. Thomson*

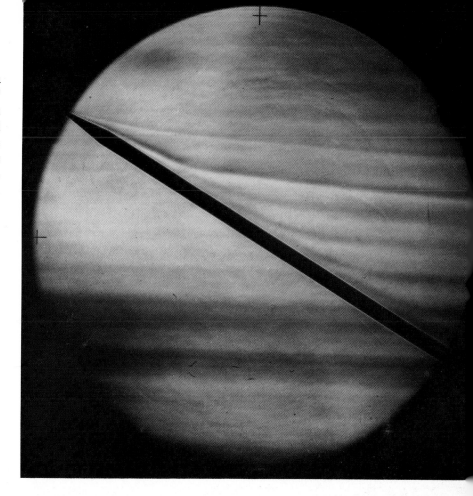

89. Vortices shedding behind a body in supersonic flow. Here the angle of attack is 35° and the free-stream Mach number is 1.6, so that the component of Mach number normal to the axis of the body is 0.92. The bow shock wave therefore moves progressively farther ahead of the cylinder downstream. A weak shock wave springs from the cone-cylinder juncture. Other shock waves appear between the rear of the body and the series of vortices that are shed alternately. *Photograph by K. D. Thomson*

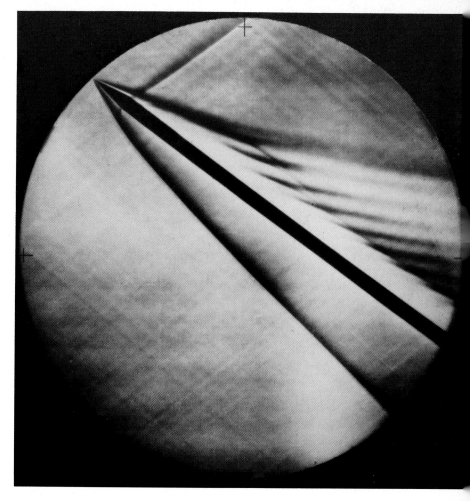

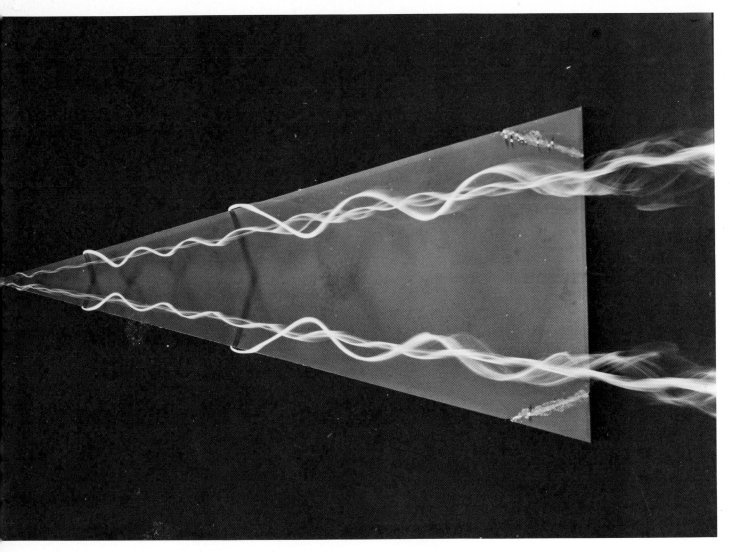

90. Vortices above an inclined triangular wing. Lines of colored fluid in water show the symmetrical pair of vortices behind a thin wing of 15° semi-vertex angle at 20° angle of attack. The Reynolds number is 20,000 based on chord. Although the Mach number is very low, the flow field is practically conical over most of the wing, quantities being constant along rays from the apex. *ONERA photograph, Werlé 1963*

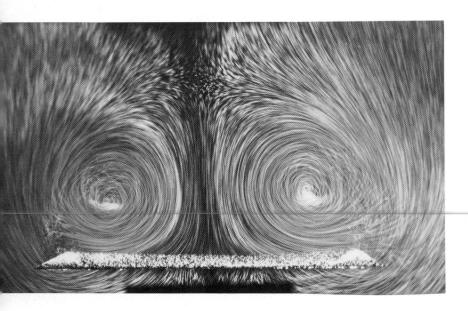

91. Cross section of vortices on a triangular wing. Tiny air bubbles in water show the vortex pair for the flow above in a section at the trailing edge of the wing. *ONERA photograph, Werlé 1963*

92. Horseshoe vortices ahead of a cylinder in a boundary layer. The laminar boundary layer on a flat plate separates ahead of a short circular cylinder, whose height is about three times the boundary-layer thickness. The vorticity in the boundary layer concentrates into three vortices that wrap around the front of the cylinder. Closer to the plate, two vortices of opposite sign form in the reverse flow, and are reflected in the plate. The Reynolds number is 5000 based on cylinder diameter. Visuali-zation is by smoke filaments in air, illuminated by a thin slice of light in the symmetry plane. This shows three stagnation points on the plate, three points of attachment, and two free stagnation points between the vortices. Another picture of the same flow appears as the frontispiece to Thwaites' *Incompressible Aerodynamics. Photograph from E. P. Sutton and the Cambridge University Engineering Department.*

93. Horseshoe vortices ahead of a cylinder in a boundary layer. In this plan view the thickness of the on-coming Blasius boundary layer is one-third of the diameter of the cylinder, as in the photograph above, and the Reynolds number is 4000 based on the diameter, but the cylinder is two diameters rather than half a diameter high. The horseshoe vortices are made visible by a sheet of smoke introduced into the boundary layer upstream. *Photograph by Sadatoshi Taneda*

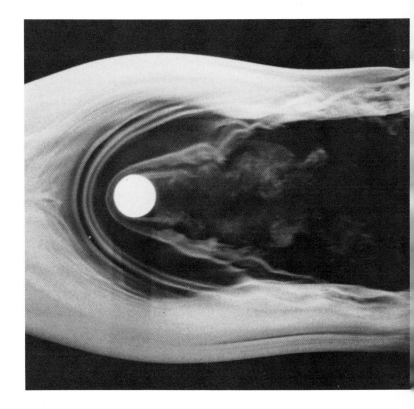

94. Kármán vortex street behind a circular cylinder at R=140. Water is flowing at 1.4 cm/s past a cylinder of diameter 1 cm. Integrated streaklines are shown by electrolytic precipitation of a white colloidal smoke, illuminated by a sheet of light. The vortex sheet is seen to grow in width downstream for some diameters. *Photograph by Sadatoshi Taneda*

95. Kármán vortex street behind a circular cylinder at R=200. This photograph, made using a different fluid (and in another country) happens to have been timed so as to resemble remarkably the flow pattern in the upper picture. A thin sheet of tobacco smoke is introduced upstream in a low-turbulence wind tunnel. *Photograph by Gary Koopmann*

96. Kármán vortex street behind a circular cylinder at $R=105$. The initially spreading wake shown opposite develops into the two parallel rows of staggered vortices that von Kármán's inviscid theory shows to be stable when the ratio of width to streamwise spacing is 0.28. Streaklines are shown by electrolytic precipitation in water. *Photograph by Sadatoshi Taneda*

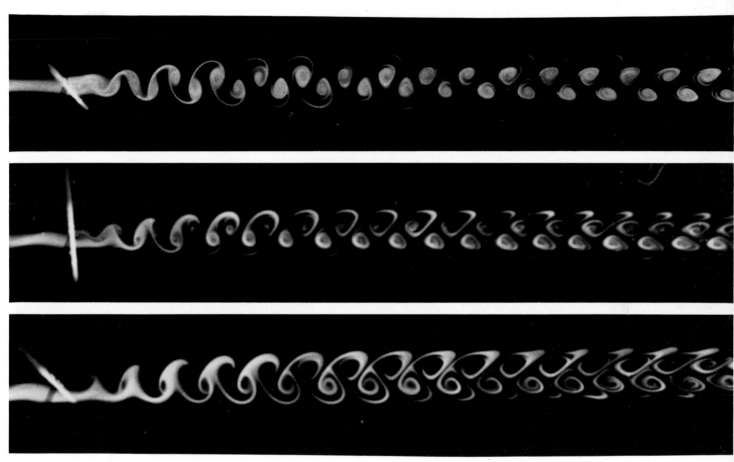

97. Smoke at various levels in a vortex street. A smoke filament in air shows, at a Reynolds number of 100, both shear layers (top photographs), only one shear layer (middle), and the irrotational flow below the wake (bottom). *Zdravkovich 1969*

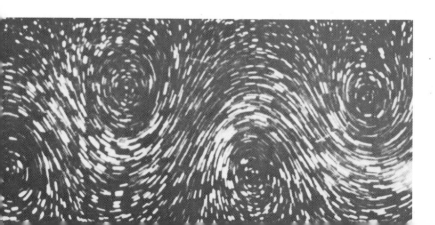

98. Kármán vortices in absolute motion. Here the camera moves with the vortices rather than the cylinder. The streamline pattern closely resembles the inviscid one calculated by von Kármán. The flow is visualized by particles floating on water. *Photograph by R. Wille, from Werlé 1973. Reproduced, with permission, from the* Annual Review of Fluid Mechanics, *Volume 5,* © *1973 by Annual Reviews Inc.*

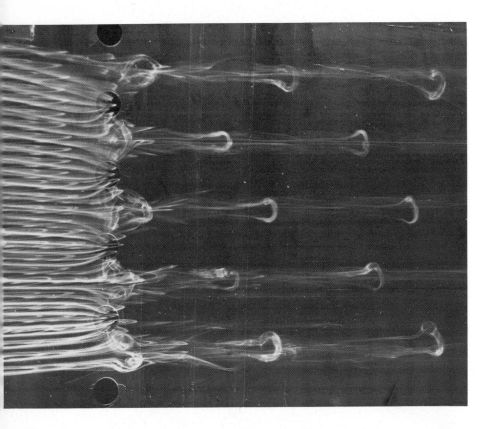

99. Plan view of horseshoe vortices in a laminar boundary layer. A laminar boundary layer in a smoke wind tunnel is seen flowing across a row of holes to which suction is applied. At lower suction rates a pair of counter-rotating vortices extends straight downstream from each hole, and the flow remains steady. Here, with higher suction, horseshoe-shaped vortex loops detach periodically from the surface and are convected downstream. *Photograph from Peter Bradshaw and Aerodynamics Division, National Physical Laboratory*

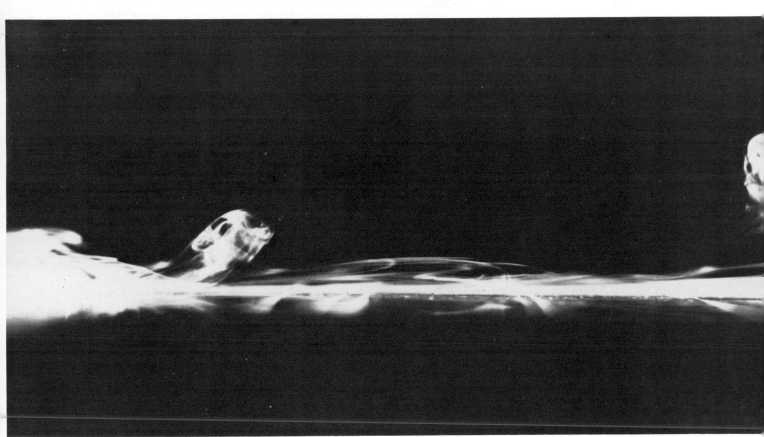

100. Side view of horseshoe vortices in a laminar boundary layer. From the side, the vortex loops in the photograph above are seen to rise well outside the boun- dary layer as they move downstream. *Photograph from Peter Bradshaw and Aerodynamics Division, National Physical Laboratory*

101. Inertial waves on a bathtub vortex. In this demonstration at the San Francisco Exploratorium, devised by Doug Hollis, water is injected tangentially at the top of the tank and drawn off at the bottom, at a rate being controlled by the girl at the right. The air-core vortex, familiar from a bathtub, shows a varicose surface as a result of inertial waves. *Photograph by Nancy Rodger*

5. Instability

102. Instability of an axisymmetric jet. A laminar stream of air flows from a circular tube at Reynolds number 10,000 and is made visible by a smoke wire. The edge of the jet develops axisymmetric oscillations, rolls up into vortex rings, and then abruptly becomes turbulent. *Photograph by Robert Drubka and Hassan Nagib*

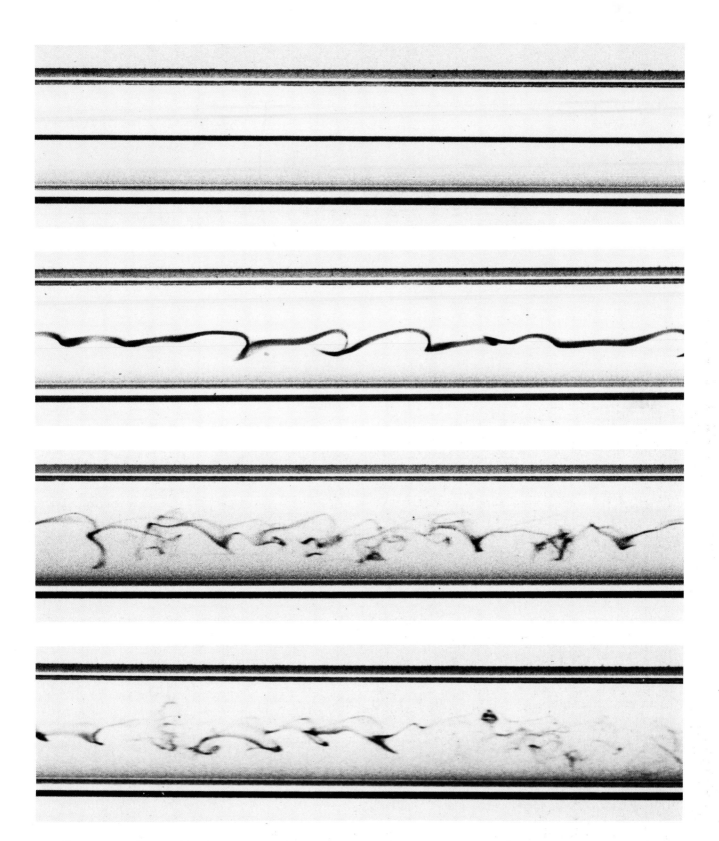

103. Repetition of Reynolds' dye experiment. Osborne Reynolds' celebrated 1883 investigation of stability of flow in a tube was documented by sketches rather than photography. However the original apparatus has survived at the University of Manchester. Using it a century later, N. H. Johannesen and C. Lowe have taken this sequence of photographs. In laminar flow a filament of colored water introduced at a bell-shaped entry extends undisturbed the whole length of the glass tube. Transition is seen in the second of the photographs as the speed is increased; and the last two photographs show fully turbulent flow. Modern traffic in the streets of Manchester made the critical Reynolds number lower than the value 13,000 found by Reynolds.

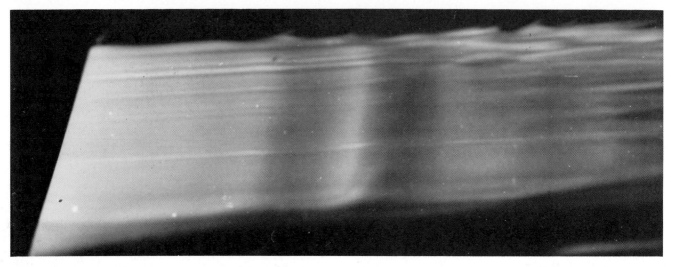

104. Instability of the boundary layer on a plate. At $R = 20,000$ based on length (upper photograph) the boundary layer is laminar over a flat plate aligned with the stream. At $R = 100,000$ (lower photograph) two-dimensional Tollmien-Schlichting waves appear. They are made visible by colored fluid in water. *ONERA photographs, Werlé 1980*

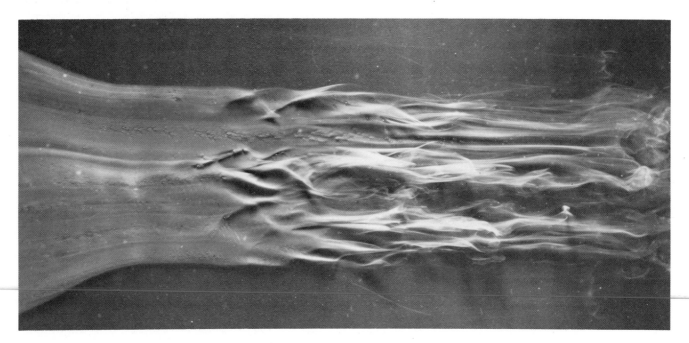

105. Natural transition on a slightly inclined plate. At the same Reynolds number of 100,000 but 1° angle of attack, transition to turbulence occurs on the plate. *ONERA photograph, Werlé 1980*

106. Transition downstream of Tollmien-Schlichting waves. The waves are two-dimensional at the left, becoming three-dimensional as they roll up in the middle, and turbulent at the right. Streaklines are shown in this perspective view by dye in water at a Reynolds number of 400,000 based on length. *Wortmann 1977*

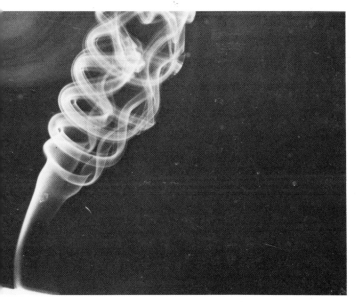

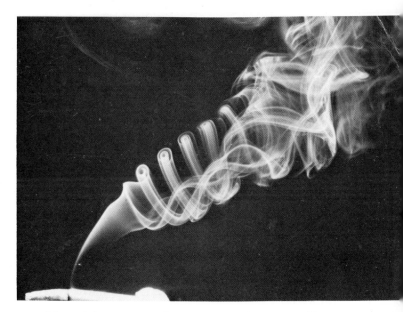

107. Instability of smoke from a cigarette. These two views of a familiar phenomenon, photographed in a slight draft, show that the initially rectilinear plume becomes unstable and forms two trailing vortices interconnected by a ladder-like network of vortex loops. *Perry & Lim 1978*

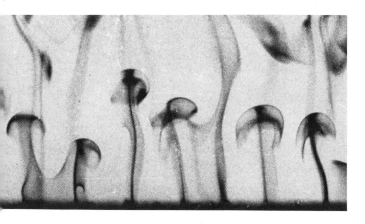

108. Buoyant thermals rising from a heated surface. Mushroom-shaped plumes rise periodically above a heated copper plate. They are made visible by an electrochemical technique using thymol blue. The heating rate is higher in the photograph at the right. *Sparrow, Husar & Goldstein 1970*

109. Emmons turbulent spot. On a flat plate, transition from a laminar to a turbulent boundary layer proceeds intermittently through the spontaneous random appearance of spots of turbulence. Each spot grows approximately linearly with distance while moving downstream at a fraction of the free-stream speed, and maintaining the characteristic arrowhead shape that is shown here by a suspension of aluminum flakes in water. Transverse contamination is seen spreading from the bottom of the channel. At the center of the spot the Reynolds number is 200,000 based on distance from the leading edge. *Cantwell, Coles & Dimotakis 1978*

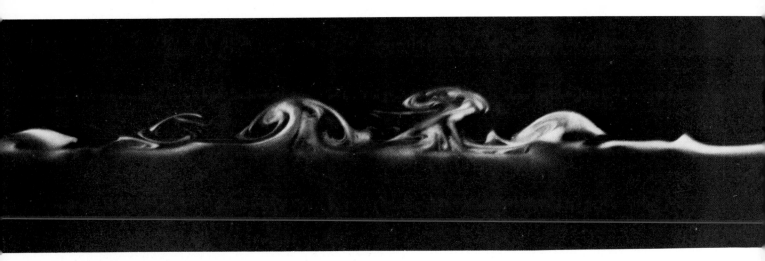

110. Cross section of a turbulent spot. A turbulent spot at an early stage of development is seen in a cross section normal to the stream. Smoke in a wind tunnel is illuminated by a sheet of laser light. *Perry, Lim & Teh 1981*

R=100,000

R=200,000

R=400,000

111. Turbulent spot at different Reynolds numbers.
The outline of the spot becomes more regular, and the
angle of its leading edge steeper, as the Reynolds number
increases. Visualization is by smoke in air with flood
lighting. *Photograph by R. E. Falco*

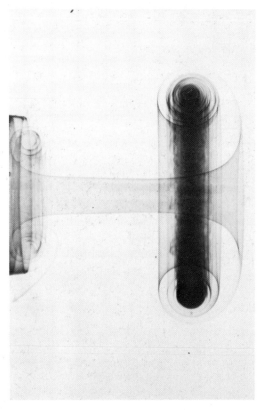
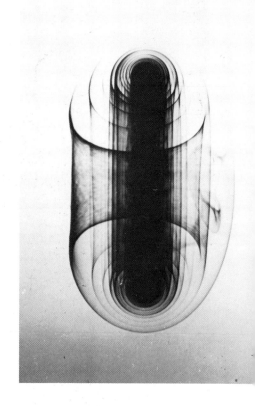
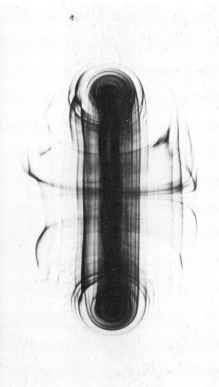

112. Instability of a laminar vortex ring. The top row of photographs shows ejection of dye-laden water from a 5-cm orifice producing an axisymmetric vortex ring similar to the one in air in figure 77. Its Reynolds number is about 1500. The lower photographs show its subsequent destruc- tion by instability. Sinusoidal perturbations develop with seven waves around the ring. The outer layers are seen to be distorted opposite to the core. The waves grow in am- plitude until the ring abruptly undergoes a transition to turbulence, with its structure still visible. *Didden 1977*

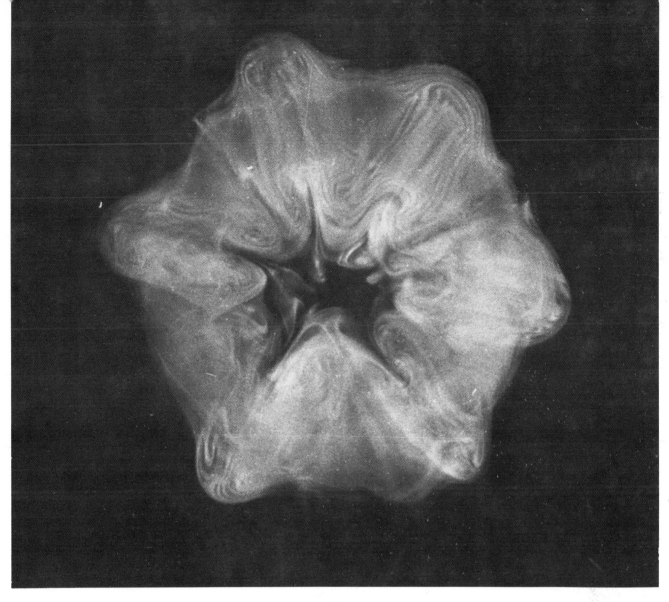

113. Hexagonal smoke ring. The growth of waves around a vortex ring is often called Widnall instability, after the researcher who first analyzed it. Here it has pro- duced a remarkably symmetric pattern of smoke in air at a Reynolds number of about 1000. *Photograph by G. J. Jameson & M. Urbicain*

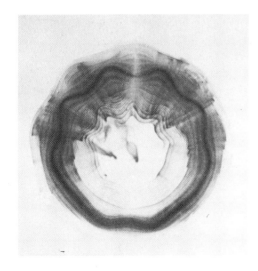 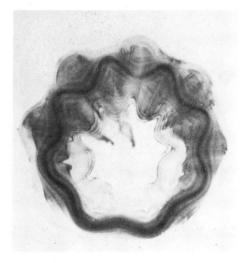 

114. Growth of waves on a vortex ring. At a higher Reynolds number, about 2000, this sequence shows eight waves growing on a laminar vortex ring in water. *Didden 1977*

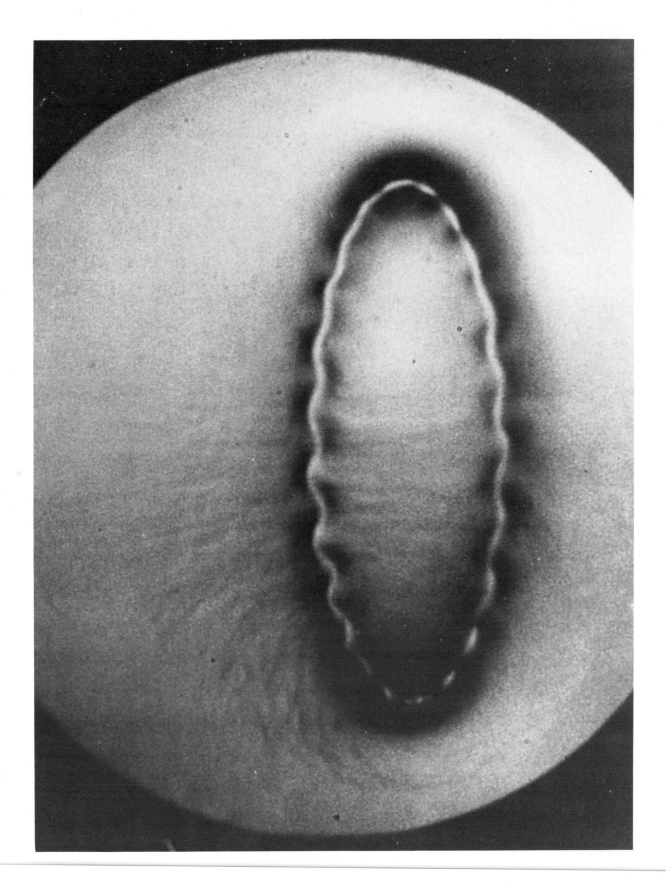

115. Unstable vortex ring at high Reynolds number. The number of waves around the vortex ring increases with Reynolds number. Here a ring in air, seen in a 15-cm field of view, is visualized by schlieren photography. It was produced by a shock tube of 8-cm diameter located 50 cm to the left, giving a Reynolds number of about 40,000. *Photograph by Bradford Sturtevant*

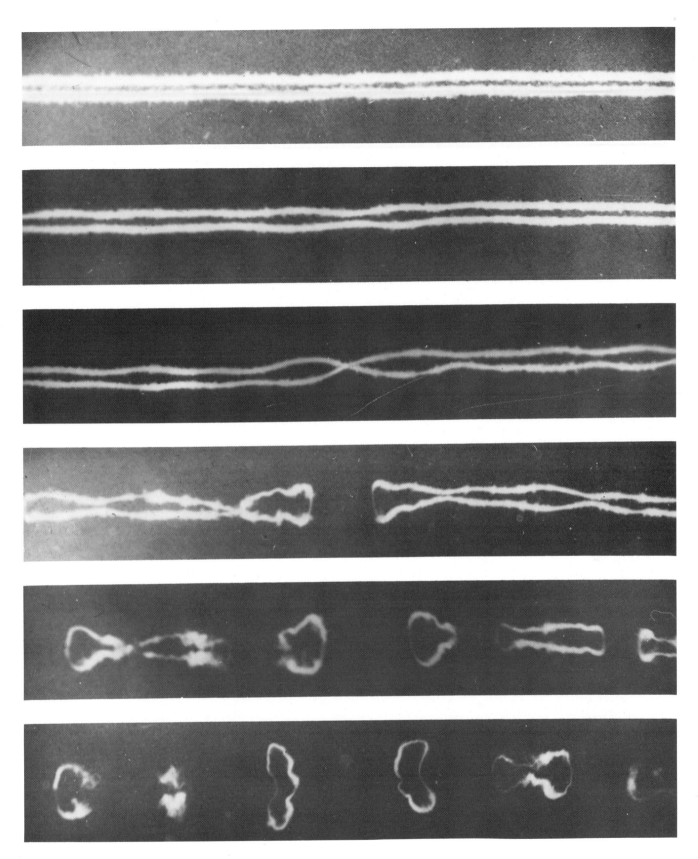

116. Instability of a pair of trailing vortices. The vortex trail of a B-47 aircraft was photographed directly overhead at intervals of 15 s after its passage. The vortex cores are made visible by condensation of moisture. They slowly recede and draw together in a symmetrical nearly sinu- soidal pattern until they connect to form a train of vortex rings. The wake then quickly disintegrates. This is commonly called Crow instability after the researcher who explained its early stages analytically. *Crow 1970, courtesy of Meteorology Research Inc.*

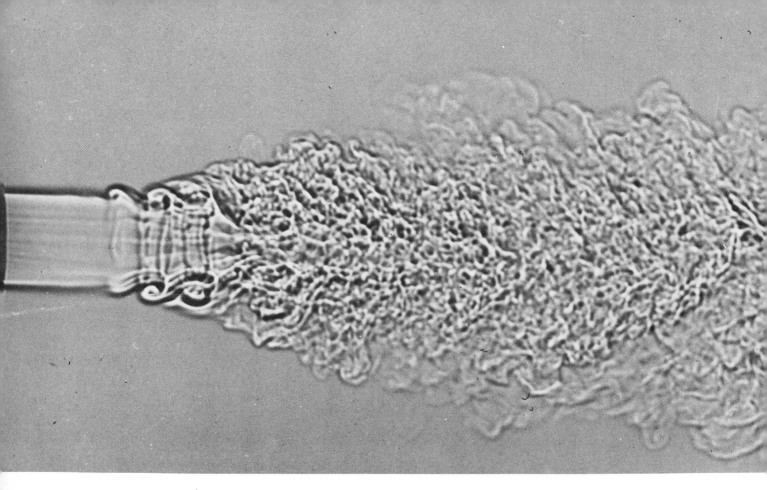

117. Instability of a round jet. This shadowgraph shows a ¼-inch jet of carbon dioxide issuing into air at a speed of 127 ft/s. It is laminar as it leaves the nozzle at a Reynolds number of approximately 30,000. One diameter downstream it shows instability, formation of vortex rings, and transition to turbulence. *Photograph by Fred Landis and Ascher H. Shapiro*

118. Instability of a round jet. Smoke gives a different view of the flow above, at a Reynolds number of about 13,000. The wavy instability of the vortex rings and their subsequent breakdown is similar to that in figure 114. *Photograph by R. Wille and A. Michalke, courtesy of H. Fiedler*

119. Unstable laminar jet impinging on a plate. The shear layer of a jet is made visible by dye in water at a Reynolds number of 4000 based on diameter and exit speed. A flat plate is placed three diameters from the nozzle. The development of the jet is modified by the feedback from the impinging vortices. *Photograph by Ho, Chih-Ming*

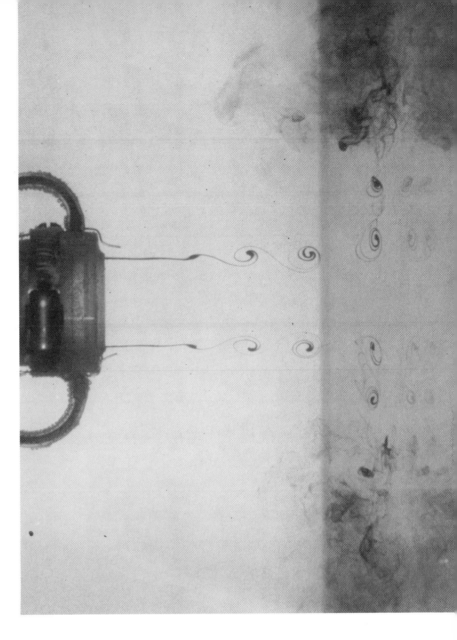

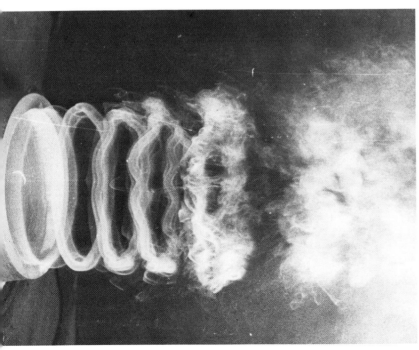

120. Forced instability of a round jet. Here weak periodic sound waves are introduced through a loudspeaker near the jet at its natural frequency. This reduces the length of the laminar boundary layer on the periphery of the jet, and causes more regular formation of vortex rings than under the unforced conditions of the figure opposite. *Photograph by R. Wille and A. Michalke, courtesy of H. Fiedler*

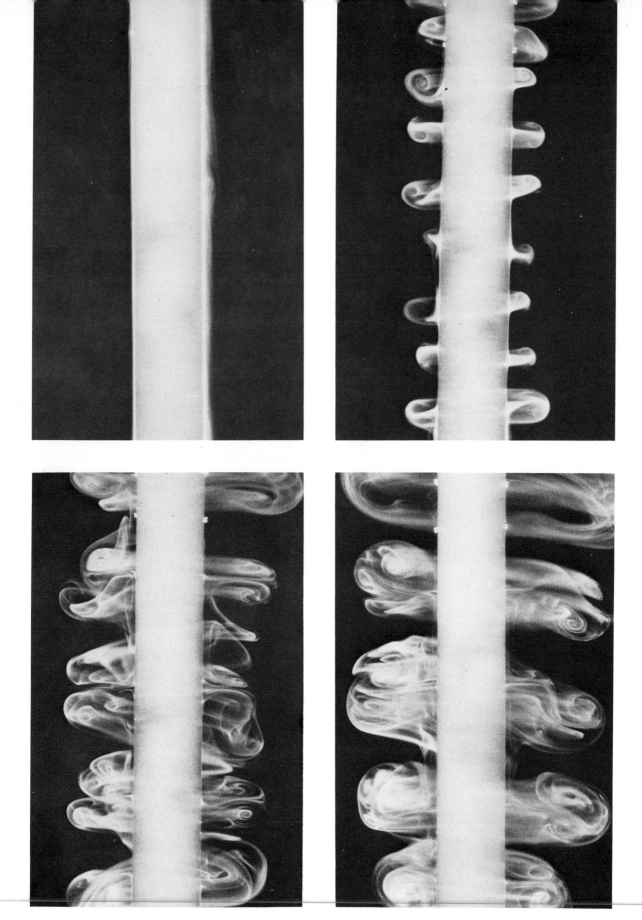

121. Growth of the boundary layer on an impulsively rotated cylinder. A 1-cm rod is started rotating in still water. The Reynolds number is 109 based on surface speed. The electrolytic precipitation method shows a uniform boundary layer after 0.6 seconds. Ring-shaped vortices have formed at 4.7 s, and are seen to grow and merge at 8.7 s and 12.3 s. *Taneda 1977*

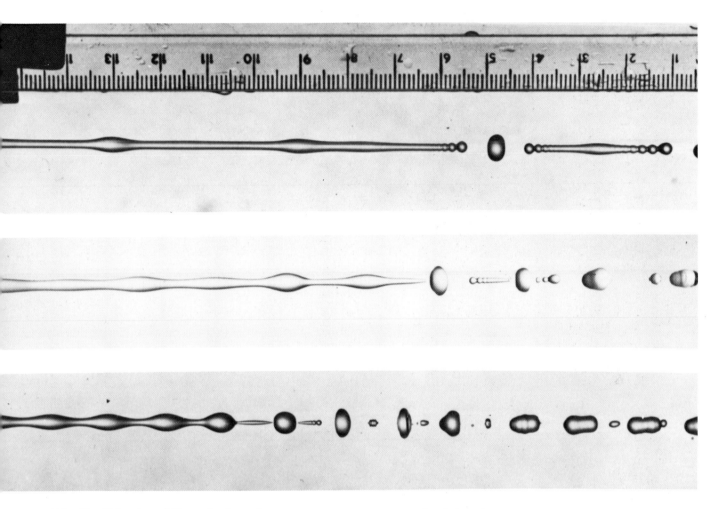

122. Capillary instability of a liquid jet. Water forced from a 4-mm tube is perturbed at various frequencies by a loudspeaker. The wavelength is 42, 12.5, and 4.6 diameters, the last being nearly Rayleigh's value for maximum growth of disturbances. The top two photographs show secondary swellings between the primary crests. *Rutland & Jameson 1971*

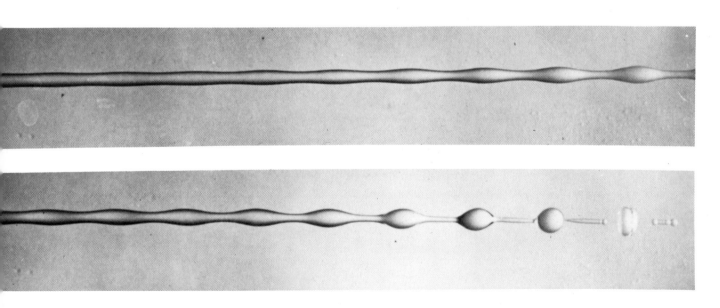

123. Effect of rotation on capillary instability. The instability of the upper jet of water, with excitation at wavelength 4.8 diameters, is increased in the lower photograph by rotating it at 435 revolutions per minute. *Rutland & Jameson 1970*

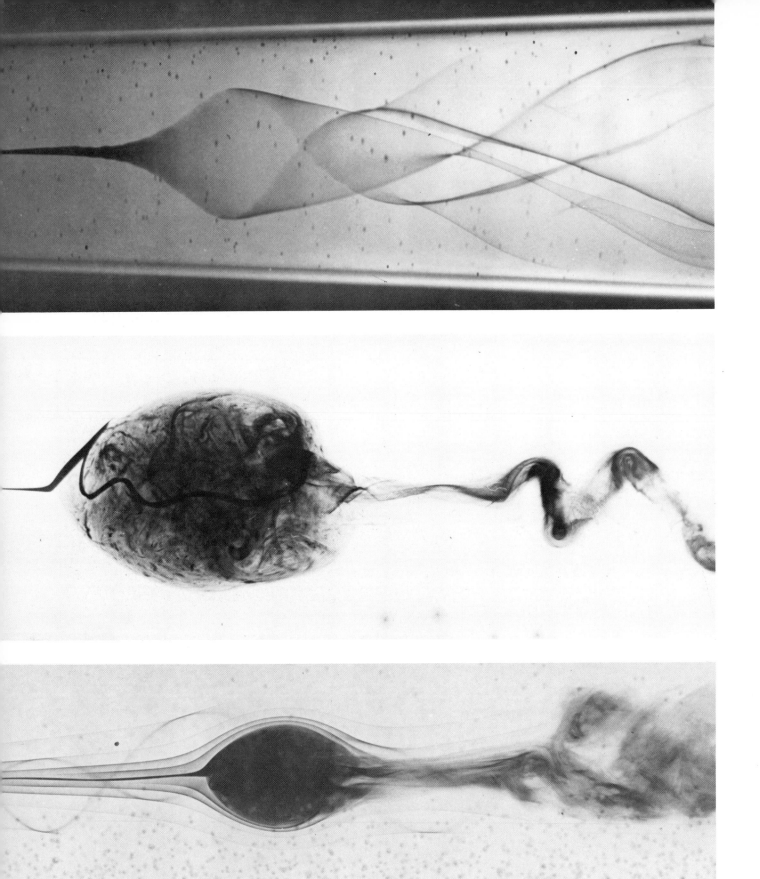

124. Vortex breakdown in swirling flow through a tube. Water to which swirl has been imparted by vanes upstream flows through a slightly diverging tube. Dye is injected upstream. Three basic types of vortex breakdown are observed. A double-helix pattern (top) forms below about $R=2000$. At higher speeds (middle) the vortex core spirals through a standing "bubble" of recirculating fluid. With still greater swirl (bottom) a smooth, almost axisymmetric body of fluid is formed. *Sarpkaya 1971*

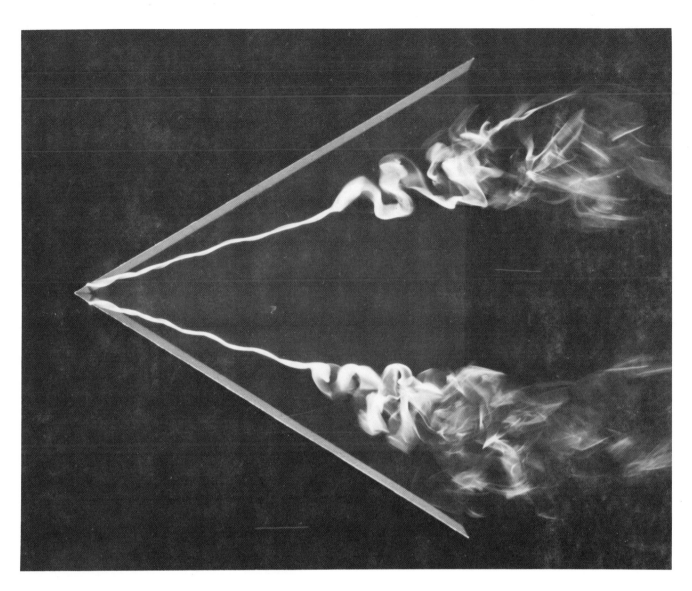

125. Vortex breakdown above a triangular wing. A thin wing of equilateral planform is seen from above at 20° angle of attack in a water tunnel. The Reynolds number is 5000 based on the chord of 10 cm. Filaments of colored fluid show that the pair of laminar vortices that roll up from separation at the leading edges abruptly burst into pockets of turbulent fluid. *ONERA photograph, Werlé 1960a*

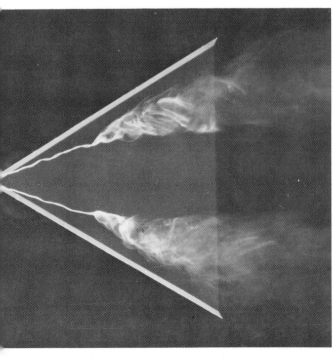

126. Effect of Reynolds number on vortex breakdown. As the Reynolds number is increased from 5000 in the photograph above to 10,000 here, the vortex breakdown moves upstream. This is almost its limiting position, unaltered by further increase of speed up to $R=20,000$. *ONERA photograph, Werlé 1960a*

127. Axisymmetric laminar Taylor vortices. Machine oil containing aluminum powder fills the gap between a fixed outer glass cylinder and a rotating inner metal one, of relative radius 0.727. The top and bottom plates are fixed. The rotation speed is 9.1 times that at which Taylor predicts the onset of the regularly spaced toroidal vortices seen here. The flow is radially inward on the heavier dark horizontal rings and outward on the finer ones. The motion was started impulsively, giving narrower vortices than would result from a smooth start. *Burkhalter & Koschmieder 1974*

128. Laminar Taylor vortices in a narrow gap. A larger inner cylinder in the apparatus to the right gives a radius ratio of 0.896. Again only the inner cylinder rotates. The upper photograph shows the center section of axisymmetric vortices at 1.16 times the critical speed. In the lower, at 8.5 times the critical speed, the flow is doubly periodic, with six waves around the circumference, drifting with the rotation. *Koschmieder 1979*

76

129. Taylor vortices between spheres. With a radius ratio of 0.95, the outer sphere fixed, and the inner one rotating at $R=7600$, aluminum particles in silicone oil show laminar vortices near the equator. *Sawatzki & Zierep 1970*

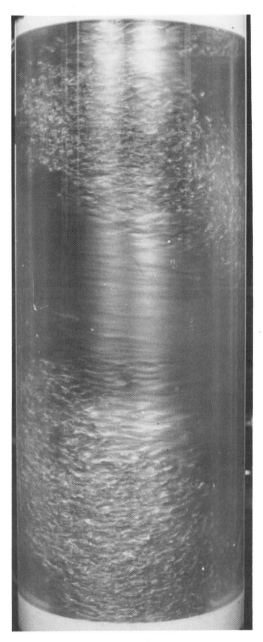

130. Spiral turbulence between counter-rotating cylinders. This "barber-pole" pattern of alternating laminar and turbulent spirals, a phenomenon discovered by Coles, was formed by first rotating the outer cylinder from rest to $R=10,000$ (based on outer radius) and then accelerating the inner one slowly to $R=4200$ (based on inner radius). *Photograph by M. Gorman and H. L. Swinney*

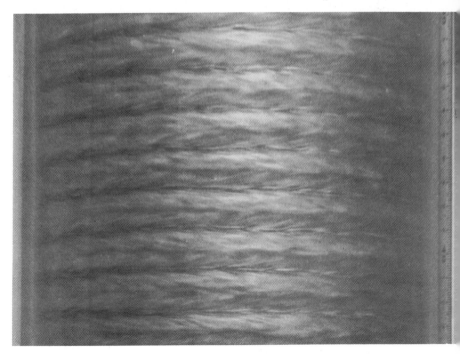

131. Axisymmetric turbulent Taylor vortices. The conditions are as in the pair of photographs on the opposite page, but at 1625 times the critical speed. A sudden start produces chaotic motion at first, but this regular permanent turbulent pattern emerges within a minute. *Koschmieder 1979*

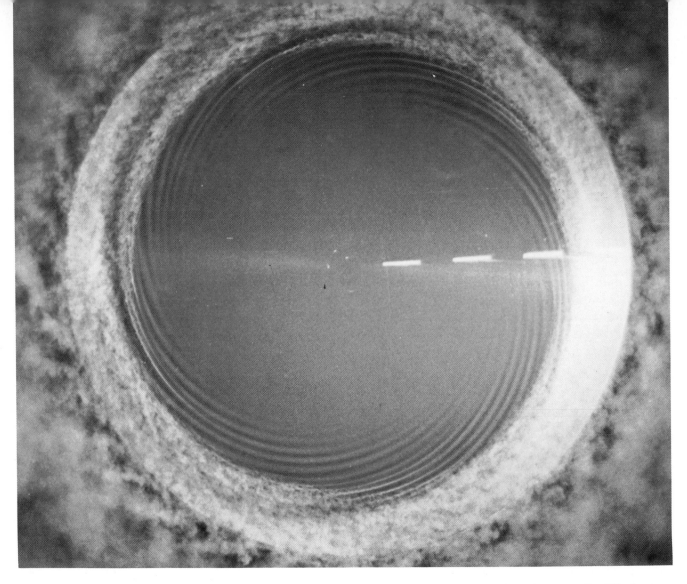

132. Spiral vortices on a spinning disk. A disk of 40-cm diameter is spinning counterclockwise in air at 1800 rpm. Gas from titanium tetrachloride applied to the black painted surface is observed stroboscopically. The laminar boundary layer in the center undergoes transition through some 32 vortices. *Kobayashi, Kohama & Takamadate 1980*

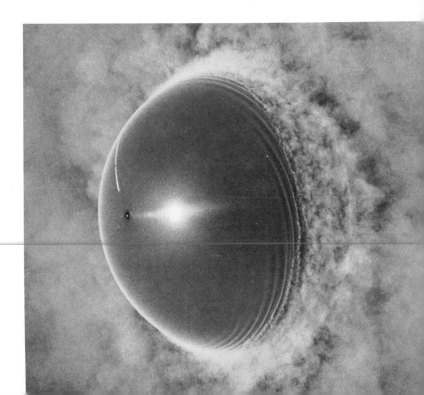

133. Spiral vortices on a spinning sphere. A sphere of 25-cm diameter is spinning in air at 1500 rpm with the rotation vector to the left. Stroboscopic observation of titanium tetrachloride gas shows that, as for the disk above, spiral vortices appear in the transition region of the boundary layer. The radial white line is used to synchronize the stroboscope. *Photograph by Y. Kohama and R. Kobayashi*

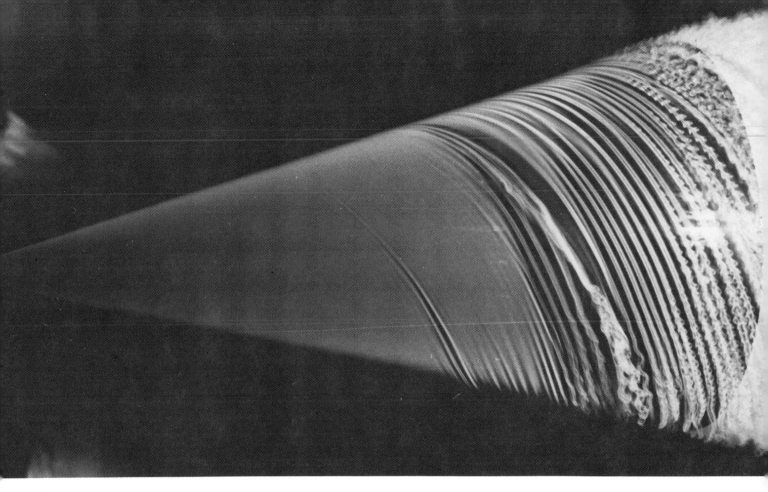

134. Spiral vortices on a cone rotating in a stream. A cone of 15° semi-vertex angle and base diameter 20 cm is spinning at 700 rpm in an air stream of 2.9 m/s. The spiral vortices are seen to assume a lacy structure as they develop. *Photograph by R. Kobayashi, Y. Kohama, and M. Kurosawa*

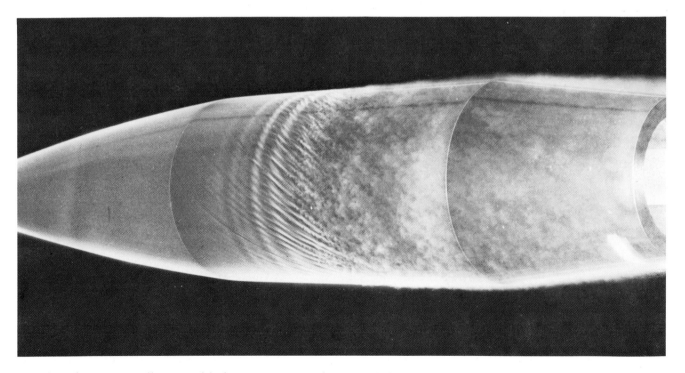

135. Simultaneous Tollmien-Schlichting waves and spiral vortices. Two modes of instability are seen superimposed on a spinning body in a smoke wind tunnel. The Reynolds number is about one million based on length, and the spin velocity at the surface of the cylindrical section is 0.61 of the free-stream speed. *Mueller, Nelson, Kegelman & Morkovin 1981*

136. Convection in a rotating cylinder. The flow pattern in a rotating "dishpan" heated at the rim and cooled at the center bears a remarkable resemblance to the basic features of the middle-latitude atmosphere of the Earth. Here water 4 cm deep in a cylinder 19.5 cm in diameter is rotating counterclockwise at 30 revolutions per second, and heated a nominal 150 W at the rim wall. A streak photograph of aluminum particles on the free surface shows narrow jets and a broad spectrum of irregular vortices. The general structure resembles a hemispheric weather map for the upper troposphere. *Fultz et al. 1959*

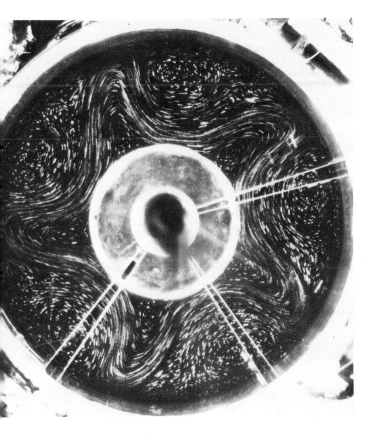

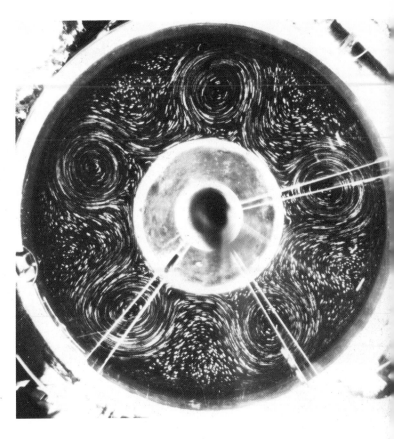

137. Convection in a rotating annulus. The wave and jet motions opposite become much more regular and periodic when rotating fluid is confined to an annulus by a central core. Symmetrical motions are observed outside a stability boundary while steady waves or periodically fluctuating waves occupy much of the inside region. Within the wave region, as rotation increases, the waves change to successively greater wave numbers. These photographs show two phases of a spontaneous periodic fluctuation (period=16 rev.) in a five-wave system for water 7 cm deep between cylinders of radii 19.5 and 7.8 cm rotating at 0.50 sec⁻¹. Heating is a nominal 80 W at the rim and 15 W from a base ring. The waves are propagating slowly counterclockwise relative to the container. *Fultz & Spence 1967*

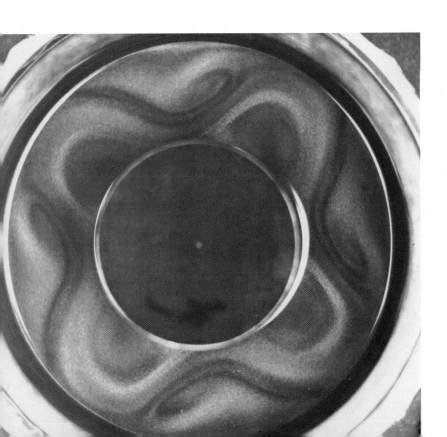

138. Convection in a rotating annulus with negative radial temperature gradient. Here the core is 15° warmer than the rim. Four waves formed by baroclinic instability are shown by aluminum powder floating in the water. The rotation is counterclockwise. The annulus has diameters of 5 and 10 cm, and the water is 10 cm deep. The rotation rate is 2.5 radians per second. *Koschmieder 1972*

139. Buoyancy-driven convection rolls. Differential interferograms show side views of convective instability of silicone oil in a rectangular box of relative dimensions 10:4:1 heated from below. At the top is the classical Rayleigh-Bénard situation: uniform heating produces rolls parallel to the shorter side. In the middle photograph the temperature difference and hence the amplitude of motion increase from right to left. At the bottom, the box is rotating about a vertical axis. *Oertel & Kirchartz 1979, Oertel 1982a*

 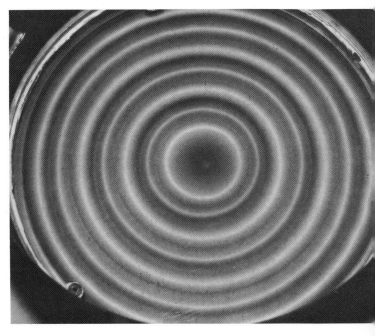

140. Circular buoyancy-driven convection cells. Silicone oil containing aluminum powder is covered by a uniformly cooled glass plate, which eliminates surface-tension effects. The circular boundary induces circular rolls. In the left photograph the copper bottom is uniformly heated at 2.9 times the critical Rayleigh number, giving regular rolls. At the right, the bottom is hotter at the rim than at the center. This induces an overall circulation which, superimposed on regular circular rolls, produces alternately larger and smaller rolls. *Koschmieder 1974, 1966*

141. Surface-tension-driven (Bénard) convection. A top view, magnified some 25 times, shows the hexagonal convection pattern in a layer of silicone oil 1 mm deep that is heated uniformly below and exposed to ambient air above. With the upper surface free, the flow is driven mainly by inhomogeneities in surface tension, rather than by buoyancy as on the previous page. Light reflected from aluminum flakes shows fluid rising at the center of each cell and descending at the edges. The exposure time is 10 s, whereas fluid moves across the cell from the center to the edge in 2 s. *Photograph by M. G. Velarde, M. Yuste, and J. Salan*

142. Imperfections in a hexagonal Bénard convection pattern. The hexagonal pattern of cells typical of convective instability driven primarily by surface tension is seen to accommodate itself to a circular boundary. Aluminum powder shows the flow in a thin layer of silicone oil of kinematic viscosity 0.5 cm²/s on a uniformly heated copper plate. A tiny dent in the plate causes the imperfection at the left, forming diamond-shaped cells. This shows how sensitive the pattern is to small irregularities. *Koschmieder 1974*

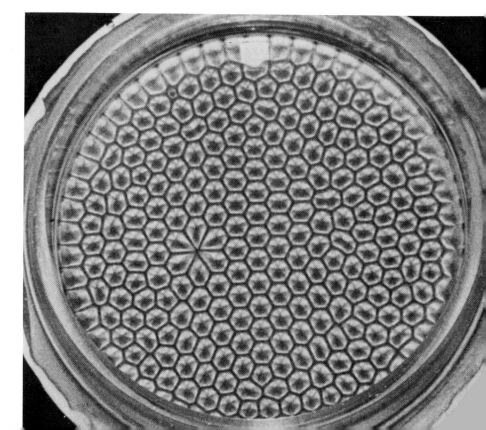

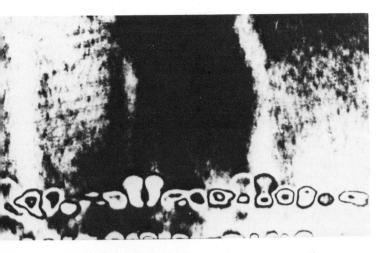
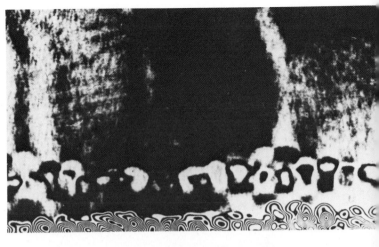
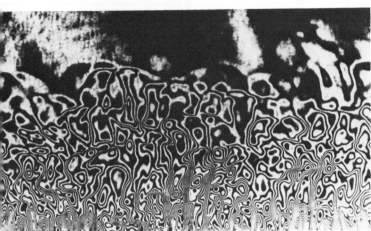
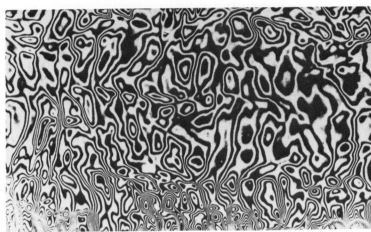

143. Instabilities in a heated layer. Freon between horizontal plates is momentarily heated from below. Successive differential iterferograms show instabilities in the starting plume, instabilities in the thermal boundary layer on the lower plate, transition to turbulence, and turbulent structures in the top and bottom boundary layers. *Oertel 1982b*

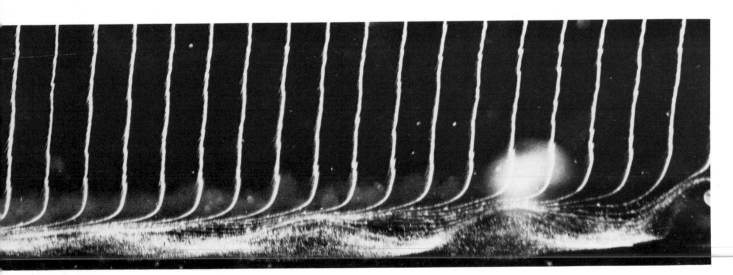

144. Secondary instability on a concave wall. A three-dimensional Tollmien-Schlichting type of instability is observed in the streamwise zones between the Taylor-Görtler vortices in an unstable boundary layer on a concave wall. The flow is from left to right, and the view along the spanwise direction. The lines of bubbles originate from a cathode wire placed perpendicular to the wall to visualize the velocity profiles. *Bippes 1972*

145. Kelvin-Helmholtz instability of stratified shear flow. A long rectangular tube, initially horizontal, is filled with water above colored brine. The fluids are allowed to diffuse for about an hour, and the tube then quickly tilted six degrees, setting the fluids into motion. The brine accelerates uniformly down the slope, while the water above similarly accelerates up the slope. Sinusoidal instability of the interface occurs after a few seconds, and has here grown nonlinearly into regular spiral rolls. *Thorpe 1971*

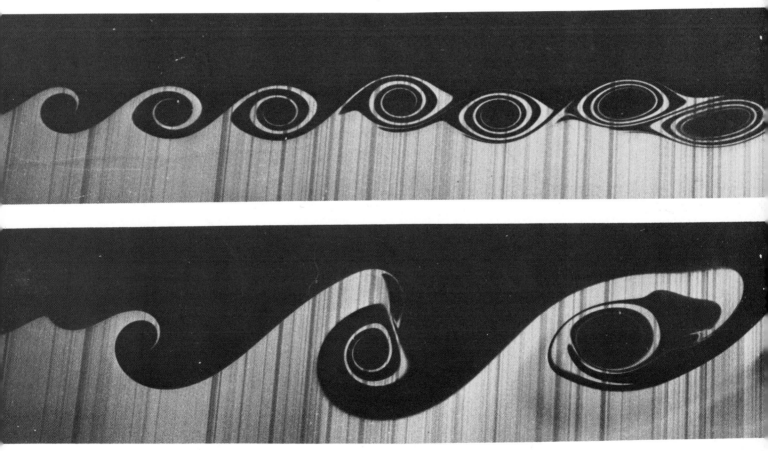

146. Kelvin-Helmholtz instability of superposed streams. The lower stream of water, moving to the left faster than the upper one, contains dye that fluoresces under illumination by a vertical sheet of laser light. The faster stream is perturbed sinusoidally at the most unstable frequency in the upper photograph, and at half that frequency in the lower one so that the motion locks into the subharmonic. *Photographs by F. A. Roberts, P. E. Dimotakis & A. Roshko*

147. Kelvin-Helmholtz waves on a thin liquid sheet. Water is ejected downward in the form of a thinning sheet from a fan-spray nozzle that is vibrated sinusoidally normal to the plane of the sheet at a frequency giving wave instability. A side view shows the initial exponential growth predicted by two-dimensional theory and the development of a complex wave form as the amplitude increases. The stream of drops along the axis at the bottom is seen in the plan view to result from collapse of the rims of the sheet. *Crapper, Dombrowski, Jepson & Pyott 1973*

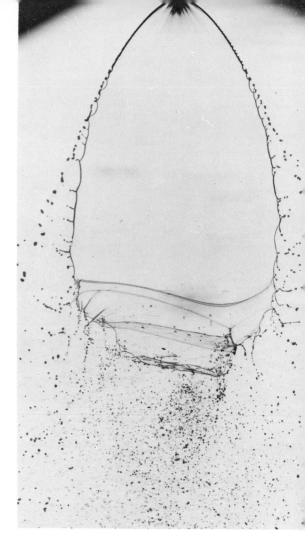

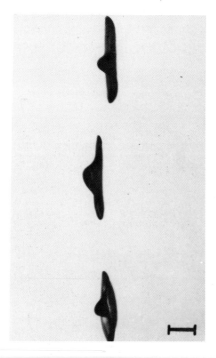

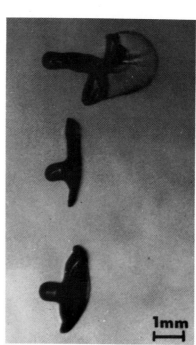

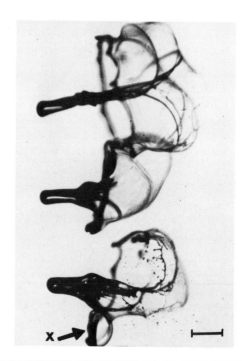

148. Breakup of droplets in an air stream. Water drops are shown at 969, 1528, and 1622 microseconds after they were injected into a stream of air blowing from left to right. The characteristic umbrella-shaped instability occurs here at a Weber number of 32.5. At only slightly lower Weber numbers it is replaced by a bag-shaped instability, evidence of which is indicated by an X in the third photograph. *Simpkins 1971. Reprinted by permission from* Nature: Physical Science, *Vol. 233, No. 37, pp. 31-33. Copyright © 1971 Macmillan Journals Limited*

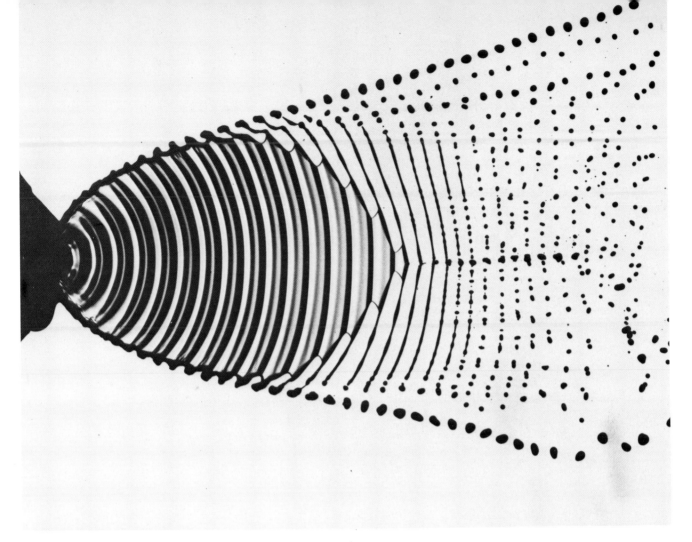

149. Breakup of a liquid sheet. Studies of drop formation in sprays show that the production of filaments generally constitutes an intermediate stage. Controlled filament formation should therefore provide a means of regulating the size of drops. One approach is to subject the nozzle to forced vibrations. This photograph shows a flat laminar sheet of water issuing from a fan-spray nozzle that is oscillated axially at resonant frequency. *Photograph by N. Dombrowski*

150. Cellular patterns in a rotating horizontal cylinder. A hollow Plexiglas cylinder partially filled with liquid rotates about a horizontal axis, the bottom moving toward the observer. Thickened fingers of fluid form at equally spaced intervals along the axis, are carried up with the rising wall, and fall back in sheets normal to the axis. *Karweit & Corrsin 1975*

6. Turbulence

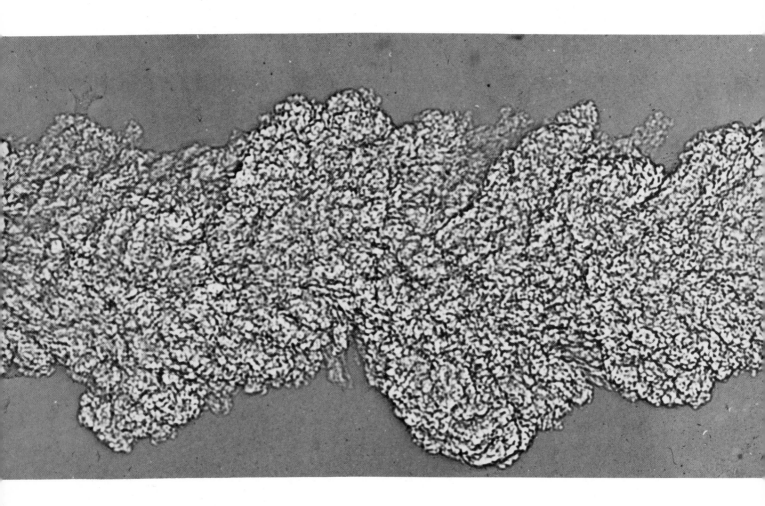

151. Turbulent wake far behind a projectile. A bullet has been shot through the atmosphere at supersonic speed, and is now several hundred wake diameters to the left. This short-duration shadowgraph shows the remarkable sharpness of the irregular boundary between the highly turbulent wake produced by the bullet and the almost quiescent air in irrotational motion outside. *Photograph made at Ballistic Research Laboratories, Aberdeen Proving Ground, in Corrsin & Kistler 1954*

152. Generation of turbulence by a grid. Smoke wires show a uniform laminar stream passing through a $\frac{1}{16}$-inch plate with ¾-inch square perforations. The Reynolds number is 1500 based on the 1-inch mesh size. Instability of the shear layers leads to turbulent flow downstream. *Photograph by Thomas Corke and Hassan Nagib*

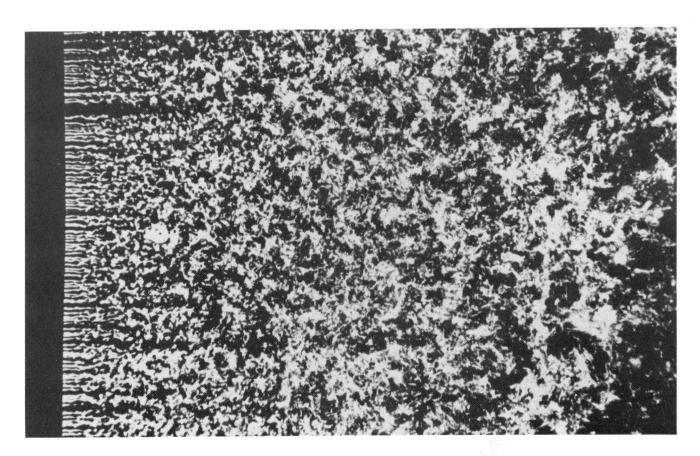

153. Homogeneous turbulence behind a grid. Behind a finer grid than above, the merging unstable wakes quickly form a homogeneous field. As it decays downstream, it provides a useful approximation to the idealization of isotropic turbulence. *Photograph by Thomas Corke and Hassan Nagib*

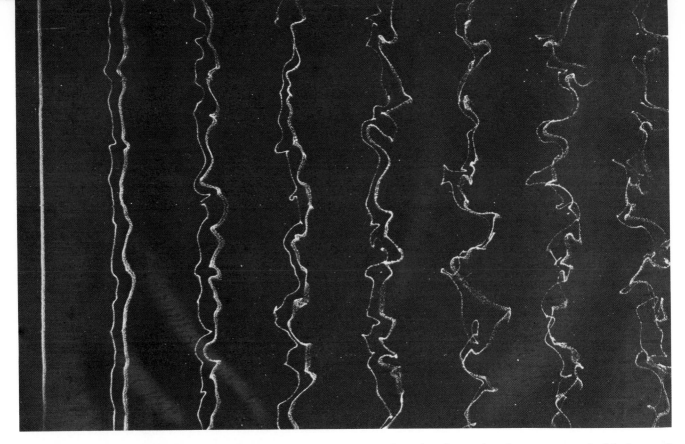

154. Growth of material lines in isotropic turbulence.
A fine platinum wire at the left is stretched across a water tunnel 18 mesh lengths behind a turbulence-generating grid. The Reynolds number is 1360 based on grid rod diam- eter. Periodic electrical pulses generate double lines of hydrogen bubbles that are stretched and wrinkled as they are convected downstream. *Corrsin & Karweit 1969*

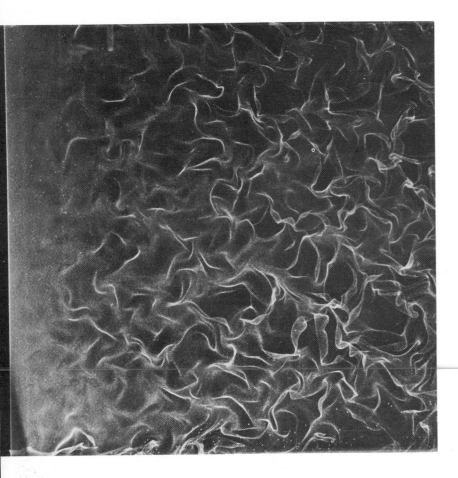

155. Wrinkling of a fluid surface in isotropic turbulence. Here the platinum wire generates a continuous sheet of hydrogen bubbles. It is deformed by the nearly isotropic turbulence behind the grid. The bright streaks are believed to be places where the crinkled sheet is viewed edge on. *Photograph by M. J. Karweit, M. S. E. thesis, Johns Hopkins Univ., 1968*

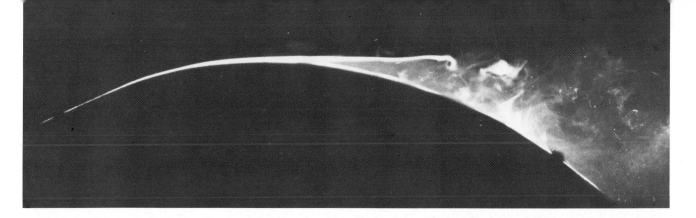

156. Comparison of laminar and turbulent boundary layers. The laminar boundary layer in the upper photograph separates from the crest of a convex surface (cf. figure 38), whereas the turbulent layer in the second photograph remains attached; similar behavior is shown below for a sharp corner. (Cf. figures 55-58 for a sphere.) Titanium tetrachloride is painted on the forepart of the model in a wind tunnel. *Head 1982*

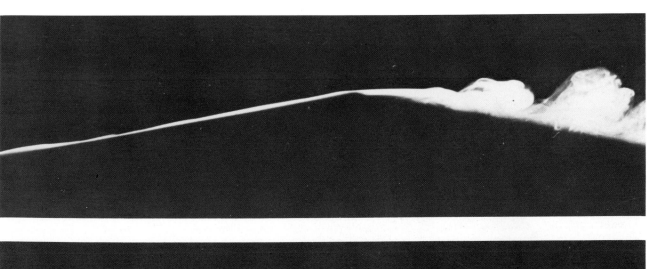

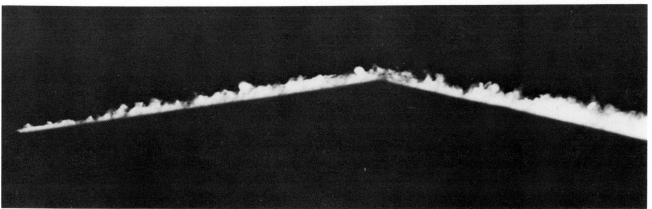

157. Side view of a turbulent boundary layer. Here a turbulent boundary layer develops naturally on a flat plate 3.3 m long suspended in a wind tunnel. Streaklines from a smoke wire near the sharp leading edge are illuminated by a vertical slice of light. The Reynolds number is 3500 based on the momentum thickness. The intermittent nature of the outer part of the layer is evident. *Photograph by Thomas Corke, Y. Guezennec, and Hassan Nagib.*

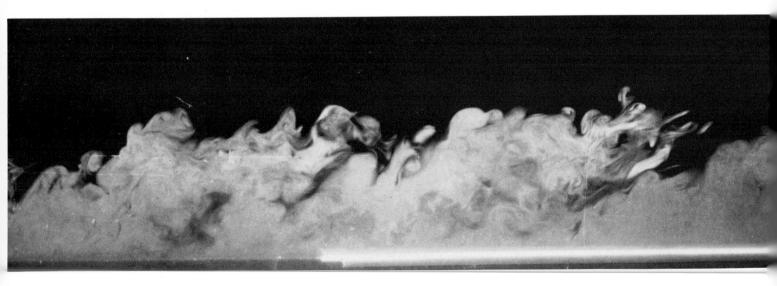

158. Turbulent boundary layer on a wall. A fog of tiny oil droplets is introduced into the laminar boundary layer on the test-section floor of a wind tunnel, and the layer then tripped to become turbulent. A vertical sheet of light shows the flow pattern 5.8 m downstream, where the Reynolds number based on momentum thickness is about 4000. *Falco 1977*

159. Sublayer of a turbulent boundary layer. A suspension of aluminum particles in a stream of water shows the streaks in the sublayer of a turbulent boundary layer on a flat wall. A mirror is used to show a simultaneous side view. *Cantwell, Coles & Dimotakis 1978*

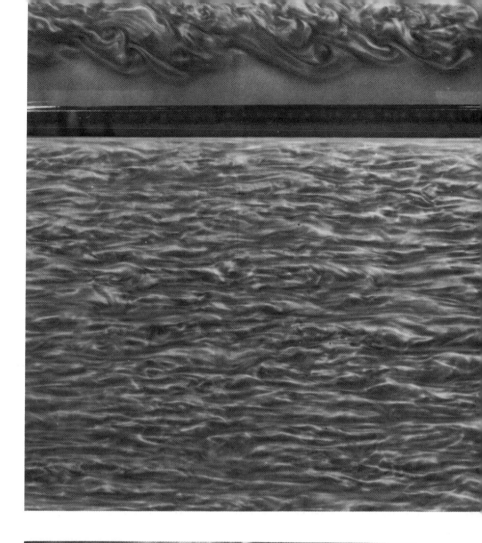

160. Detail of the sublayer. A close-up view of smoke in the turbulent boundary layer on a wind-tunnel floor shows "pockets" and streaks in the viscous sublayer. *Falco 1980*

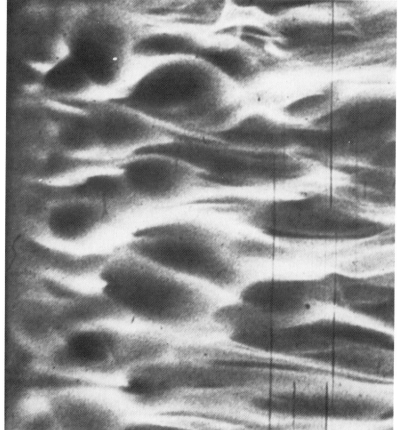

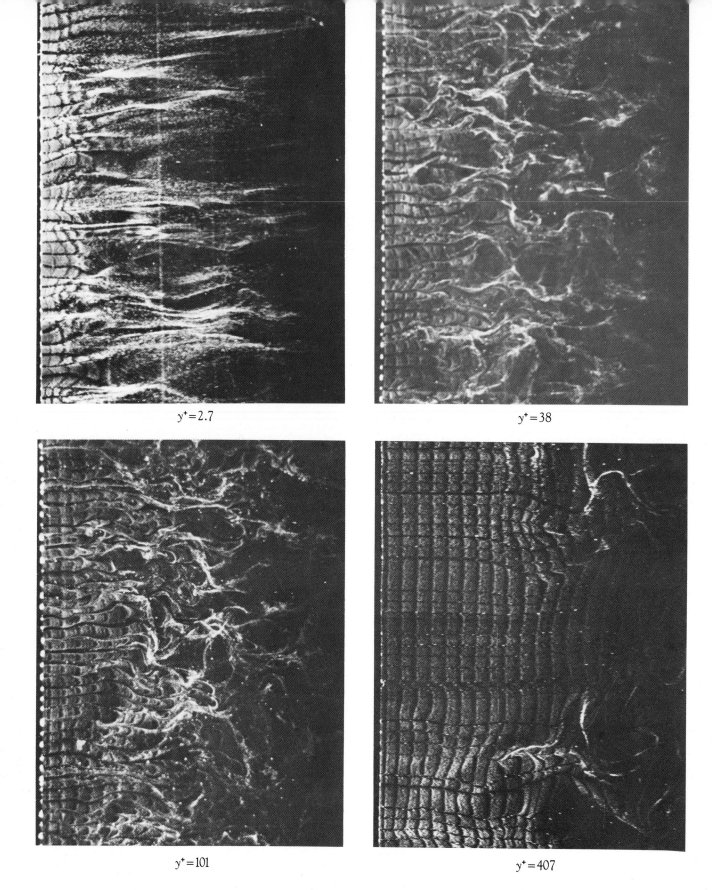

$y^+=2.7$

$y^+=38$

$y^+=101$

$y^+=407$

161. Structure of a turbulent boundary layer. Successive layers of the flow near a flat plate in a water channel are shown by tiny hydrogen bubbles released periodically from a thin platinum wire seen at the left. The height $y^+=y\,u_\tau/\nu$ of the wire above the plate is shown in wall variables, where $u_\tau=(\tau_w/\varrho)^{1/2}$ is the friction velocity. The characteristic low- and high-speed streaks shown in the viscous sublayer at $y^+=2.7$ become less noticeable farther away, and have disappeared in the logarithmic region at $y^+=101$. In the wake region at $y^+=407$ the turbulence is seen to be intermittent and of larger scale. *Kline, Reynolds, Schraub & Runstadler 1967*

162. "Typical eddy" in a turbulent boundary layer. Oil fog is illuminated by a sheet of laser light to show the lower two-thirds of a turbulent boundary layer in side view. The vortex-ring structure just below and to the right of center, which resembles a sliced mushroom leaning left, is an example of what Falcó has called a "typical eddy." It scales on wall variables (figure 161) rather than on the boundary-layer thickness. *Photograph by R. E. Falco*

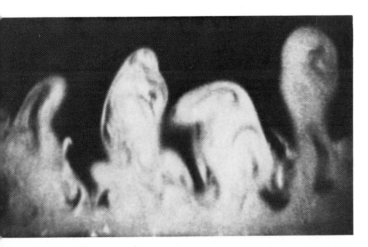
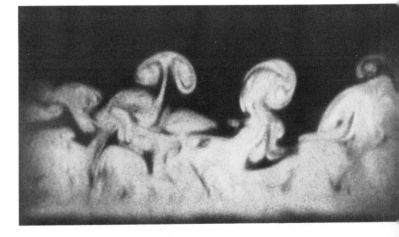

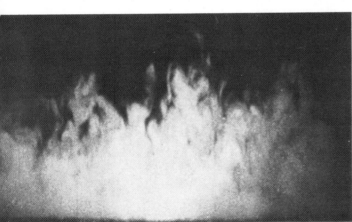

163. Oblique transverse sections of a turbulent boundary layer. The flow is viewed head-on, with smoke illuminated by a sheet of light that is inclined 45° downstream from the wall on the left and 45° upstream on the right.

The Reynolds number based on momentum thickness is 600 in the upper pair of photographs and 9400 below. *Head & Bandyopadhyay 1981*

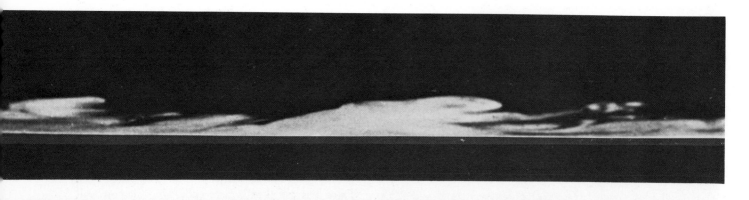

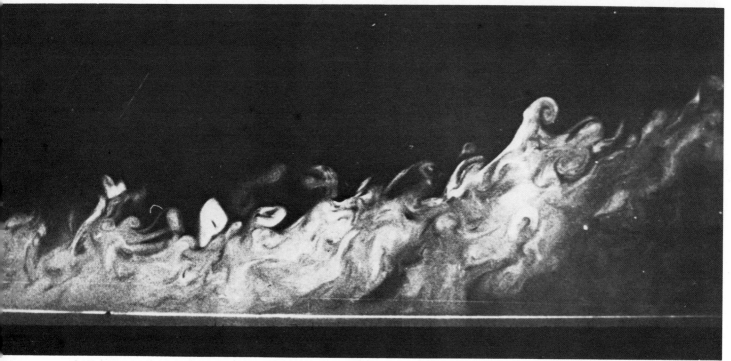

164. Effects of strong pressure gradients. In a strong
favorable pressure gradient (top) a turbulent boundary
layer is stretched and eventually relaminarizes; in a strong
adverse gradient (bottom) it thickens and separates. *Photographs by R. E. Falco, in Head 1982*

165. Turbulent boundary layer encountering a cylinder. As the turbulent boundary layer flows around a circular cylinder normal to the wall, eddies in the outer region are rapidly distorted and an intermittent reversed jet forms in the separated region. Figure 92 shows the corresponding flow with a laminar boundary layer. *Photograph by R. E. Falco*

166. Turbulent water jet. Laser-induced fluorescence shows the concentration of jet fluid in the plane of symmetry of an axisymmetric jet of water directed downward into water. The Reynolds number is approximately 2300. The spatial resolution is adequate to resolve the Kolmogorov scale in the downstream half of the photograph. *Dimotakis, Lye & Papantoniou 1981*

97

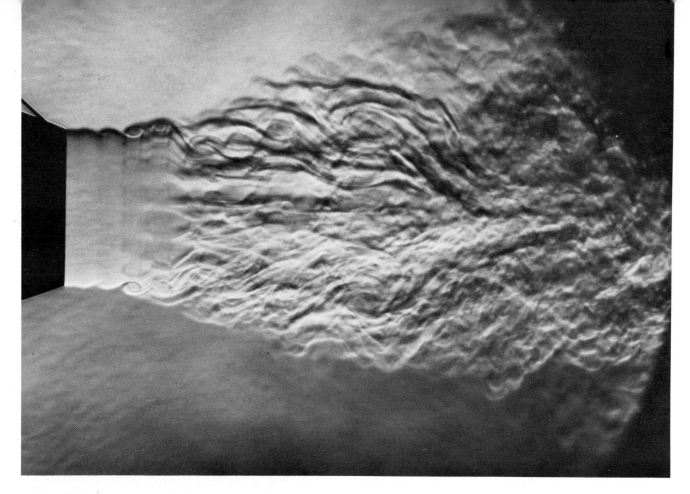

167. Subsonic jet becoming turbulent. A jet of air from a nozzle of 5-cm diameter flows into ambient air at a speed of 12 m/s. The laminar interface becomes unstable as in figure 102, and the entire jet eventually becomes turbulent. *Bradshaw, Ferriss & Johnson 1964*

26 mi/hr.

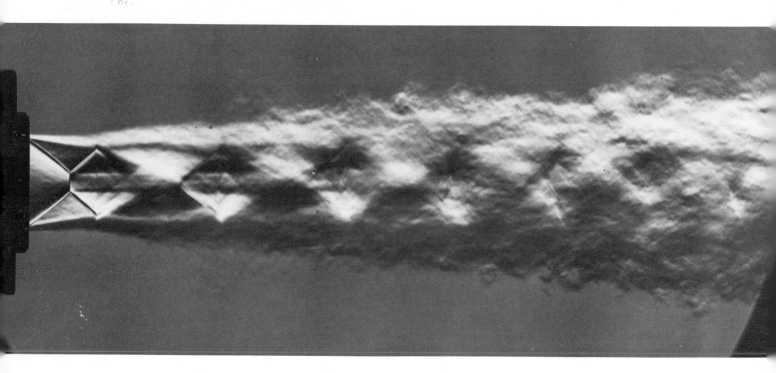

168. Supersonic jet becoming turbulent. At a Mach number of 1.8 a slightly under-expanded round jet of air adjusts to the ambient air through a succession of oblique and normal shock waves. The diamond-shaped pattern persists after the jet is turbulent. *Oertel 1975*

1370 mi/hr.

169. Entrainment by a plane turbulent jet. A time exposure shows the mean flow of a plane jet of colored water issuing into ambient water at 100 cm/s. Tiny air bubbles mark the streamlines of the slow motion induced in the surrounding water. *ONERA photograph, Werlé 1974*

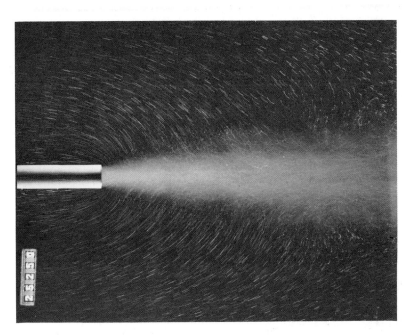

170. Entrainment by an axisymmetric turbulent jet. A jet of colored turbulent water flows from a tube of 9 mm diameter at 200 cm/s. According to boundary-layer theory the streamlines shown by air bubbles in the water outside the jet are paraboloids of revolution, and parabolas in the plane case above. *ONERA photograph, Werlé 1974*

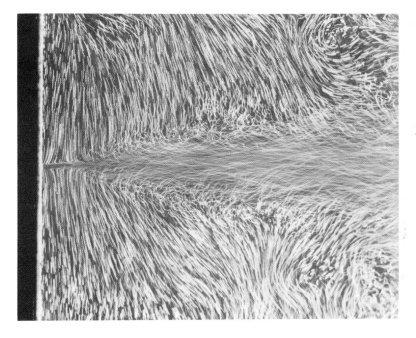

171. Plane turbulent jet from a wall. Clear water flowing at 30 cm/s from a long slit in a wall is initially laminar. After it becomes turbulent, the jet is like that in the top photograph, but the flow in the surrounding water is quite different because of the presence of the wall. *ONERA photograph, Werlé 1974*

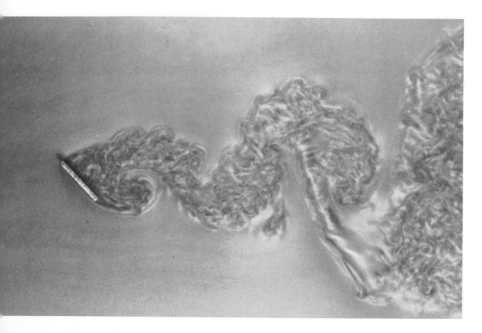

172. Wake of an inclined flat plate. The wake behind a plate at 45° angle of attack is turbulent at a Reynolds number of 4300. Aluminum flakes suspended in water show its characteristic sinuous form. *Cantwell 1981. Reproduced, with permission, from the* Annual Review of Fluid Mechanics, *Volume 13.* © *1981 by Annual Reviews Inc.*

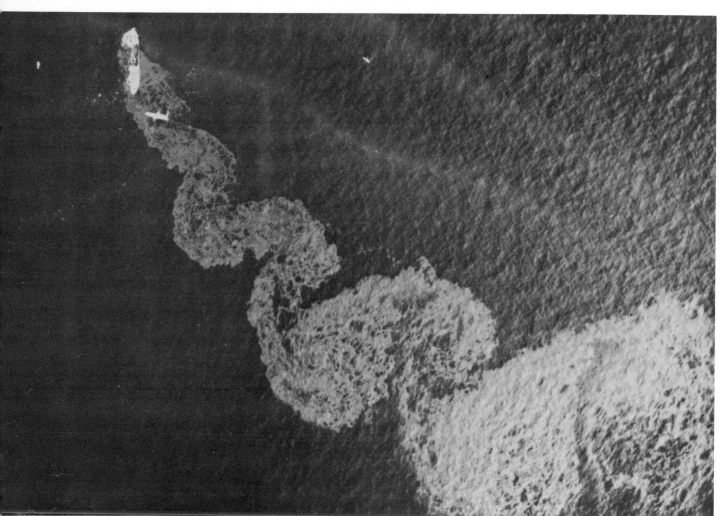

173. Wake of a grounded tankship. The tanker *Argo Merchant* went aground on the Nantucket shoals in 1976. Leaking crude oil shows that she happened to be inclined at about 45° to the current. Although the Reynolds number is approximately 10^7, the wake pattern is remarkably similar to that in the photograph at the top of the page. *NASA photograph, courtesy of O. M. Griffin, Naval Research Laboratory.*

174. Turbulent wake of a cylinder. A sheet of laser light slices through the wake of a circular cylinder at a Reynolds number of 1770. Oil fog shows the instantaneous flow pattern, covering 40 diameters centered 50 diameters downstream. *Photograph by R. E. Falco*

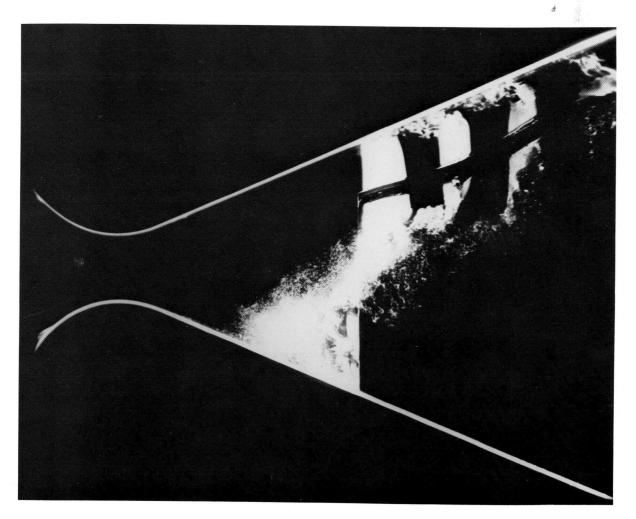

175. Separated flow in a diffuser. A pulsed hydrogen-bubble wire spanning the diffuser in water shows a turbulent boundary layer attached to the upper wall, but detachment and recirculation along the lower wall. *Kline 1963*

176. Large-scale structure in a turbulent mixing layer.
Nitrogen above flowing at 1000 cm/s mixes with a helium-argon mixture below at the same density flowing at 380 cm/s under a pressure of 4 atmospheres. Spark shadow photography shows simultaneous edge and plan views, demonstrating the spanwise organization of the large eddies. The streamwise streaks in the plan view (of which half the span is shown) correspond to a system of secondary vortex pairs oriented in the streamwise direction. Their spacing at the downstream side of the layer is larger than near the beginning. *Photograph by J. H. Konrad, Ph.D. thesis, Calif. Inst. of Tech., 1976.*

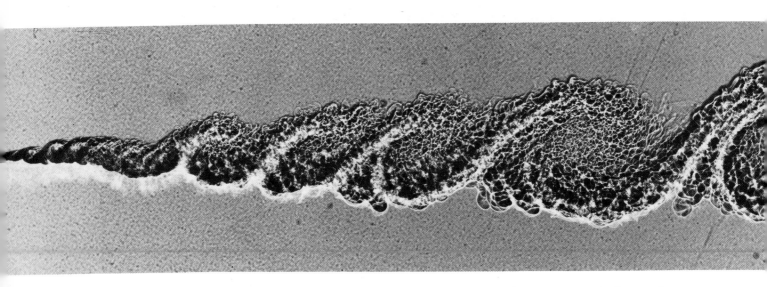

177. Coherent structure at higher Reynolds number.
This flow is as above but at twice the pressure. Doubling the Reynolds number has produced more small-scale structure without significantly altering the large-scale structure. *M. R. Rebollo, Ph.D. thesis, Calif. Inst. of Tech., 1976; Brown & Roshko 1974*

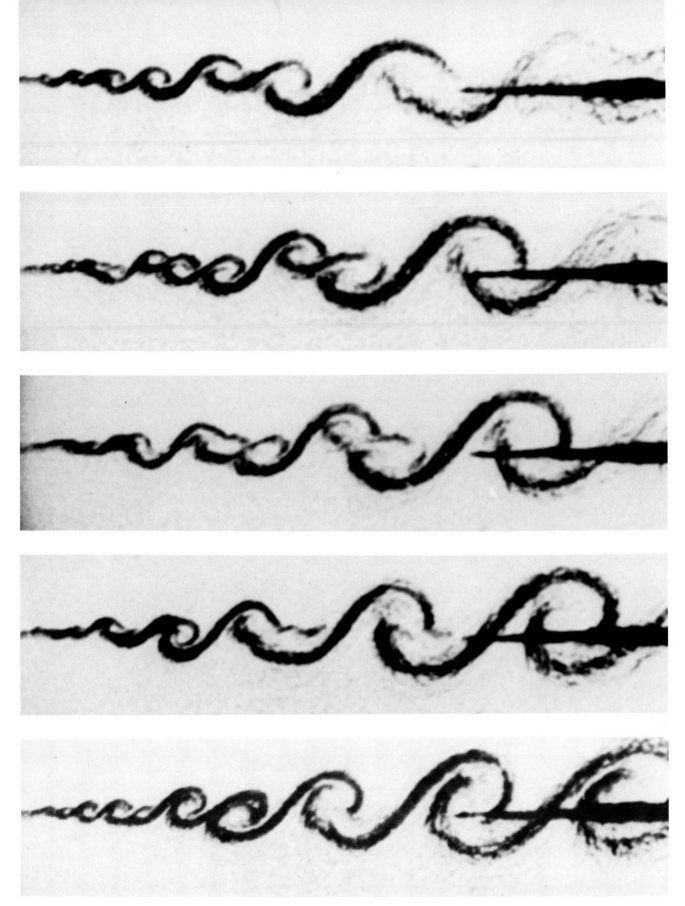

178. Pairing of vortices in a mixing layer. A sequence of shadowgraphs shows the mixing of two streams of the same density at a pressure of 8 atmospheres and a Reynolds number of 850,000. Nitrogen flows at 10 m/s above a helium-argon mixture at 3.5 m/s. A probe is visible at the right. Two vortices, distinct in the top photograph, are pairing in the center of the third one, and have become a single larger vortex at the bottom. *Photographs by L. Bernal, G. L. Brown, and A. Roshko, in Roshko 1976*

7. Free-surface Flow

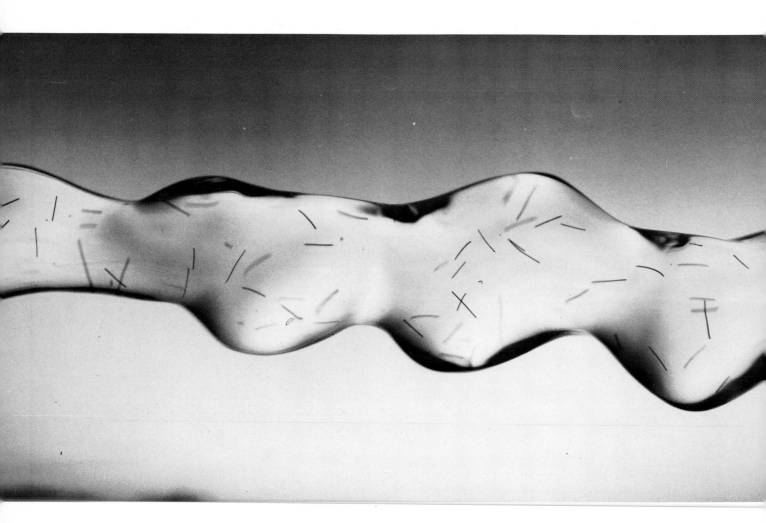

179. Water jet about to break up. Tiny rayon "floc" particles, approximately 0.015 mm in diameter and less than 1 mm long, are suspended in water as trace elements. A jet of water from a 6-mm nozzle is traveling through quiescent air at 16 m/s. In contrast to the axisymmetric mode of instability exhibited at low speeds (figure 122), the jet is destroyed by helical instability far downstream at this high Reynolds number. *Hoyt & Taylor 1982*

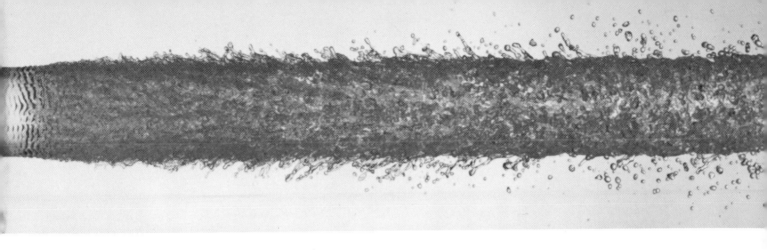

180. Water jet emerging into stagnant air. Water flows at 25 m/s from a nozzle of 6-mm diameter. The jet is seen to be initially laminar, but axisymmetric waves of insta-bility appear in the first diameter. They are amplified in a chaotic region, and culminate in the ejection of spray droplets. *Hoyt & Taylor 1977*

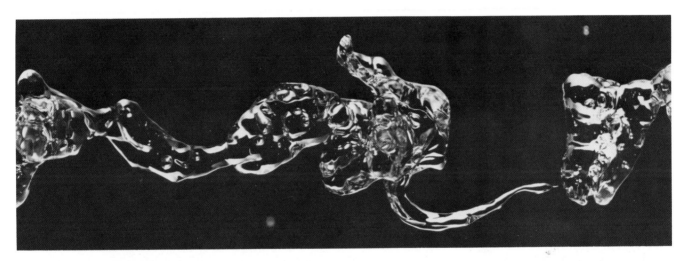

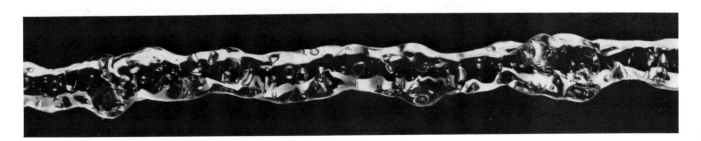

181. Water jet in a coaxial air stream. The jet is initially unaffected by flow of the surrounding air; but here 238 di-ameters downstream the helical instability is suppressed. The jet is emerging from the nozzle at 27 m/s, and the air is flowing at (top to bottom) 2, 11, and 22 m/s. *Hoyt & Taylor 1977*

182. Bubble rising in a Newtonian fluid. An air bubble of 40.6 cm³ volume rises in silicone oil along the axis of a vertical tube of 8.2 cm diameter. Effects of inertia and surface tension are neglible, the Reynolds number being less than 0.1, so that the bubble would be spherical in an unbounded medium. Here the walls make it slightly elongated, but the bubble and the flow field, shown by illuminated magnesium particles, exhibit fore-and-aft symmetry. *Coutanceau & Thizon 1981*

183. Bubble rising in a non-Newtonian fluid. A bubble of 13 cm³ volume rises at a Reynolds number less than 0.1 in a vertical tube of 8.2 cm diameter (not visible) filled with a 3.5 per cent aqueous solution of polyethylene oxide. This fluid possesses both a highly shear-thinning viscosity and considerable elasticity. The bubble loses the fore-aft symmetry it would have in a Newtonian fluid, becoming pointed at the rear. *Photograph by M. Coutanceau and P. Thizon, in van Wijngaarden & Vossers 1978*

184. Spherical-cap bubble with a laminar wake. The wake of a bubble is invisible to the naked eye or camera. Schlieren photography shows, at a Reynolds number of 90 based on the equivalent sphere radius, a laminar toroidal vortex spiralling beneath a spherical-cap bubble that is rising through a large tank of mineral oil. The bubble contains 55 cm³ of air. *Wegener & Parlange 1973. Reproduced, with permission, from the* Annual Review of Fluid Mechanics, *Volume 5. © 1973 by Annual Reviews Inc.*

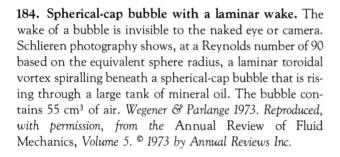

185. Spherical-cap bubbles with "skirts." Bubbles containing 50-100 cm³ of air are rising in mineral oil. For speeds above some 40 cm/s and viscosities above 200 centi-poise, a thin sheet of gas trails behind part or all of the lower edge of the bubble. *Wegener, Sundell & Parlange 1971*

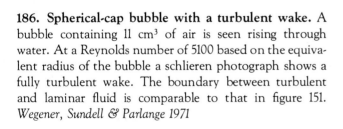

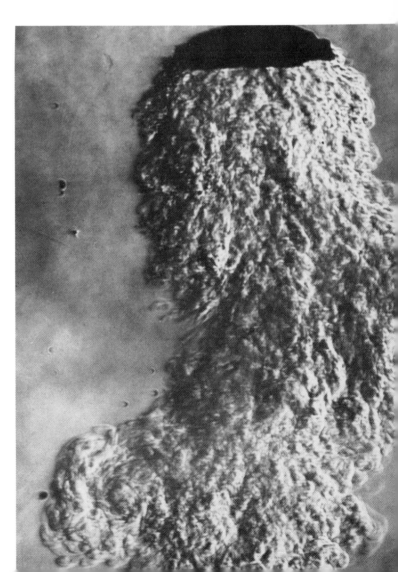

186. Spherical-cap bubble with a turbulent wake. A bubble containing 11 cm³ of air is seen rising through water. At a Reynolds number of 5100 based on the equivalent radius of the bubble a schlieren photograph shows a fully turbulent wake. The boundary between turbulent and laminar fluid is comparable to that in figure 151. *Wegener, Sundell & Parlange 1971*

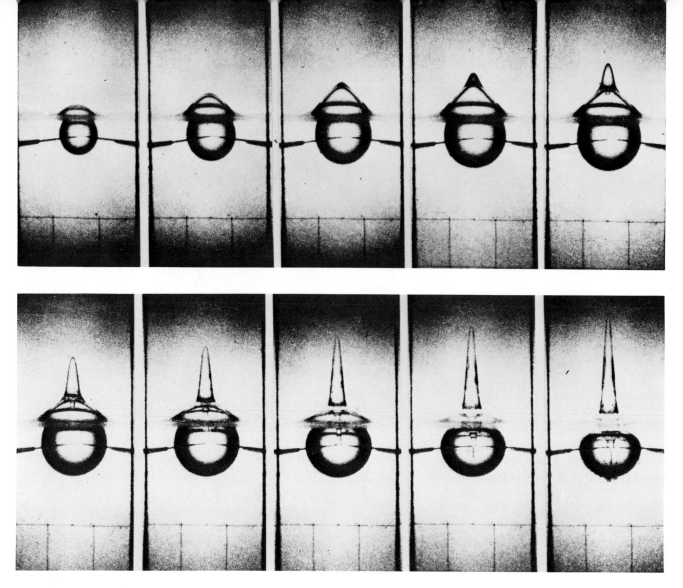

187. Collapse of a bubble near a free surface. This sequence shows the growth and collapse of a vapor bubble in water close to a free-air surface. The bubble is formed by a high-voltage spark discharge between the two probes. A spike of water penetrates the air during growth and collapse, and is balanced by a slender downward jet of water that threads the bubble from its top during the collapse. Buoyancy effects were eliminated by performing the experiment in free fall. The camera runs at about 11,000 frames per second, and the grid at the bottom is 25 mm square. *Blake & Gibson 1981*

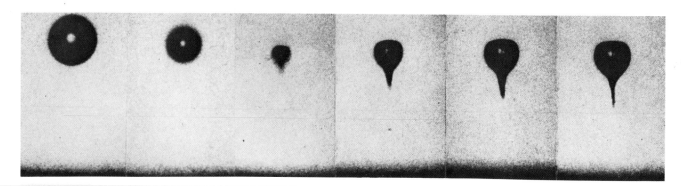

188. Collapse of a bubble near a wall. This sequence shows the collapse of a spherical bubble in still water near a plane solid surface (the dark diffuse boundary at the bottom of each frame). The bubble is produced 4.5 mm from the wall by focused ruby laser light, and has started its collapse after expanding to a maximum radius of 1.1 mm. It is photographed at the rate of 75,000 frames per second. Illumination is from behind through a ground-glass plate. The bright spot in the middle of the bubble results from light passing through undeflected. *Lauterborn 1980*

189. Jet from a bubble near a wall. The previous sequence shows the effect of a high-speed jet, directed downward, that forms at the top of the bubble during collapse. In this magnified view the jet is visible only as a thin dark vertical line through the bright spot in the middle of the bubble. Passing through the almost empty cavity, the jet impinges on its bottom and carries it along to form the spike that extends toward the wall. The jet is believed to extend far ahead of the spike, and to be the cause of cavitation erosion from a solid wall. The horizontal diameter of the bubble is about 2 mm. *Lauterborn 1980*

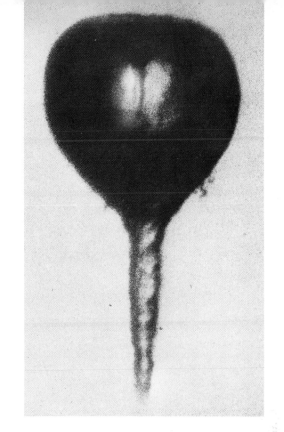

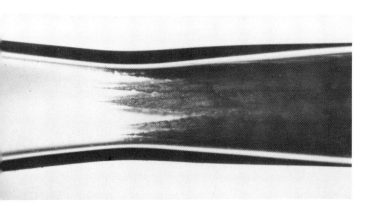

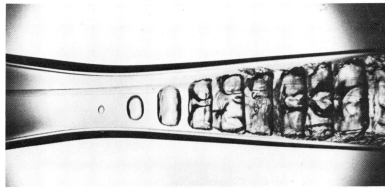

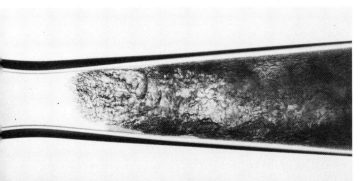

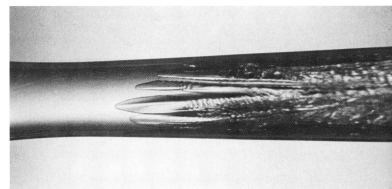

190. Cavitation in high-speed flow of water through a nozzle. Transition from simple liquid to a strongly accelerated two-phase system occurs in a nozzle throat during expansion, particularly in heated liquid. At the upper left, tap water at 204°C shows regular incipient cavitation with nuclei at the wall. The upper right shows missing nuclei, and cavitation initiated by single bubbles in the core, giving periodic pressure oscillations. At the lower right a higher fluctuation frequency is caused by retardation in boiling in water at 133°C. The lower right shows cavitation in water at room temperature containing air. *From DFG-Forschungsberichte by E. Klein, courtesy of H. Fiedler*

No reflection: pure progressive waves

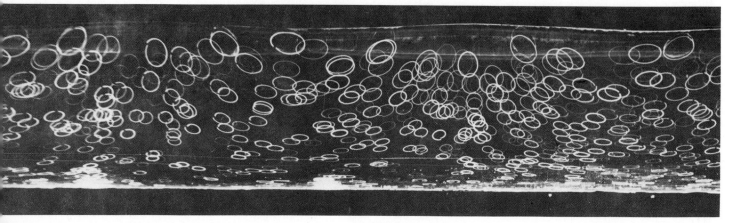

24% reflection

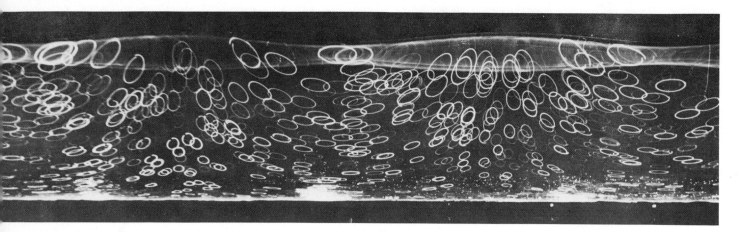

38% reflection

191. Particle trajectories in plane periodic water waves. Two wave trains of the same frequency traveling in opposite directions are produced by a progressive wave coming from the left that is reflected by a partially absorbent barrier. The top photograph shows the pure progressive wave with no reflection. Its amplitude is four per cent of the wavelength, and the water depth is 22 per cent. White particles suspended in the water are photographed during one period. Their trajectories are practically ellipses traversed clockwise, circular at the free surface and flat-tened toward the bottom. Some open loops indicate a slow drift to the right near the surface and left near the bottom. As the reflection is increased, the orbits become increas-ingly flattened and inclined. Complete reflection gives a pure standing wave in the last photograph, where the trajectories are streamlines. There the upper and lower envelopes of the water surface show that the vertical mo-tion does not vanish at the nodes. *Wallet & Ruellan 1950, courtesy of M. C. Vasseur*

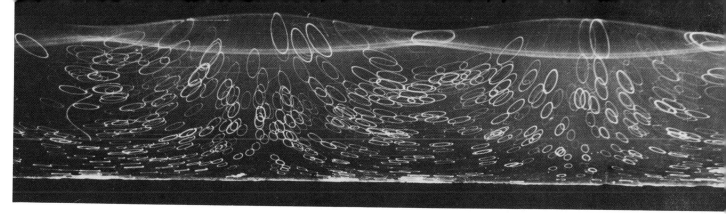

53% reflection

71% reflection

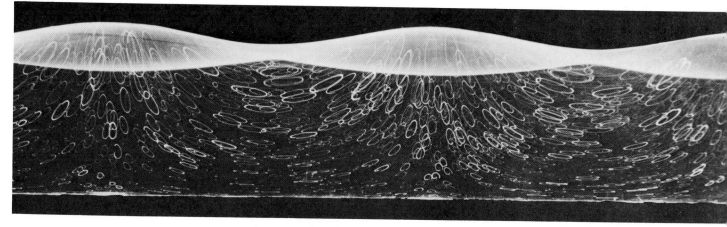

85% reflection

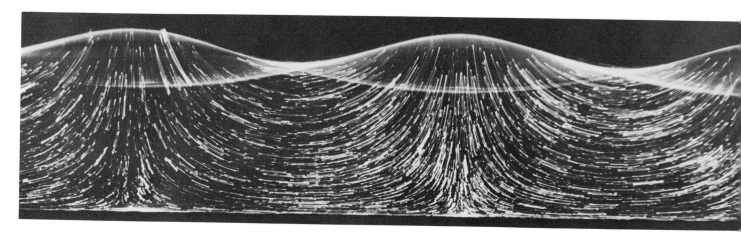

100% reflection: pure standing waves

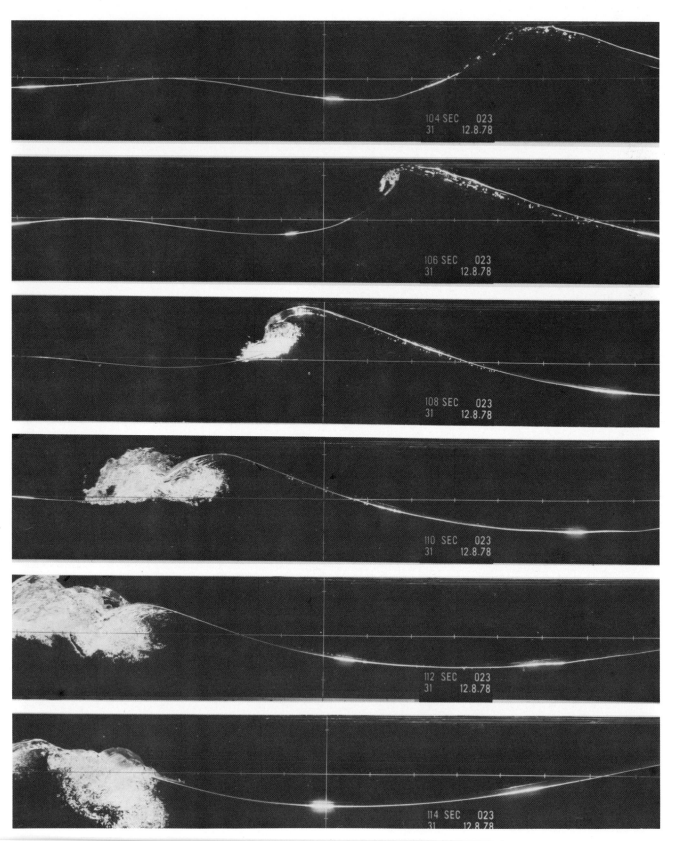

192. Breaking of a plane water wave. The wavemaker in a water tank is programmed to generate a spectrum of wavelengths, with the phases of 32 component frequencies arranged so that the crests combine when they reach the intersecting graticule lines. The wave is traveling from right to left in water 60 cm deep. The time interval between photographs is $^1/_{12}$ second (the times displayed being equivalent full-scale times). The maximum crest height is 13 cm above the mean water line. *Photographs from J. Taylor and the Edinburgh University Wave Power Project*

193. Disintegration of a train of Stokes waves. An oscillating plunger generates a train of waves, of wavelength 2.3 m in water 7.6 m deep, traveling away from the observer in a large towing basin at the Ship Division of the National Physical Laboratory. The upper photograph shows, close to the wavemaker, a regular pattern of plane waves except for small-scale roughness. In the lower photograph, some 60 m (28 wavelengths) farther along the tank, the same wave train has suffered drastic distortion. The instability was triggered by imposing on the motion of the wavemaker a slight modulation at the unstable side-band frequencies; but the same disintegration occurs naturally over a somewhat longer distance. *Photograph by J. E. Feir, in Benjamin 1967*

194. Spilling breaking waves. This regular three-dimensional pattern, reminiscent of waves in the open sea, has evolved by nonlinear instability from a uniform train of steep two-dimensional Stokes waves. The waves are propagating from left to right with wavelength 0.75 m. *Photograph by Ming-Yang Su*

195. Overall view of wave evolution. Steep Stokes waves of wavelength 0.75 m are generated in a great outdoor basin of depth 1 m by the oscillating wavemaker at the right. As they propagate to the left, instability causes them to evolve into the three-dimensional spilling breaking waves shown above in close-up. *Photograph by Ming-Yang Su*

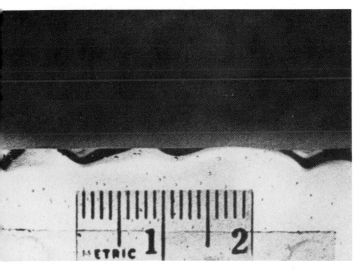

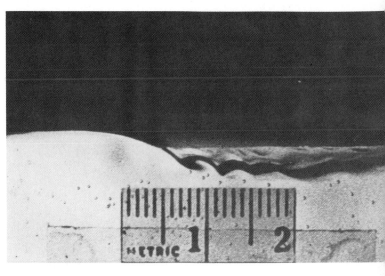

196. Capillary waves. In contrast to gravity waves, two-dimensional capillary waves may be expected theoretically to have sharp troughs and broad crests. The profiles shown were generated by a 12-knot wind blowing over a water channel. *Schooley 1958, courtesy of the Naval Research Laboratory*

197. Capillary-gravity waves. In a 16-knot wind a longer gravity wave is preceded by a train of short capillary waves. A thin film of water clings to the channel wall. The scale in these two photographs is 2 cm long. *Schooley 1958, courtesy of the Naval Research Laboratory*

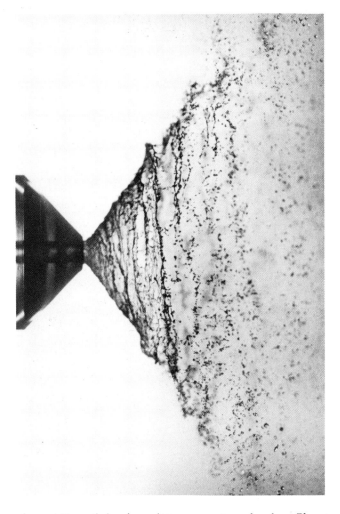

198. Atomization from a nozzle. Glycerine is ejected from a nozzle in a thin conical jet. Disturbances grow as in

figure 147 until the sheet disintegrates into droplets. *Photographs by Eugen Klein, courtesy of H. Fiedler*

199. Turbulent bore on the River Severn. When large tides travel many kilometers over shallow water the front of the tidal wave, i.e. the first part of the flood tide, steepens. It may form a "bore," which is a traveling form of hydraulic jump. The Hooghly estuary in India and that of the Tsien Tang in China have the biggest bores. Other large bores occur in the Amazon, Seine, and Severn. The Severn bore is seen here in a turbulent condition crossing water about 20 cm deep over Rodley Sands opposite Framilode. *Photograph by D. H. Peregrine*

200. Undular bore on the River Severn. The character of a bore depends on the ratio of water depths across it. If the depth of water behind the bore is less than 1.6 times the depth in front, undulations appear. These carry energy away from the bore front, and may eliminate any need for turbulent energy dissipation. This undular bore is in the River Severn at Minsterworth. The water is about 2 meters deep in mid-river (the left of the photograph) but is shallow near the bank (on the right). *Photograph by D. H. Peregrine*

201. Wave pattern of a ship. An aerial photograph from directly overhead shows, away from the breaking wave from the bow and its turbulent wake, the asymptotic pattern deduced by Kelvin in 1887. The waves are confined to an angle of 19½° on either side of the path of the ship, in agreement with the theory, with an effective origin that is displaced approximately one ship length ahead of the bow. *Newman 1970*

8. Natural Convection

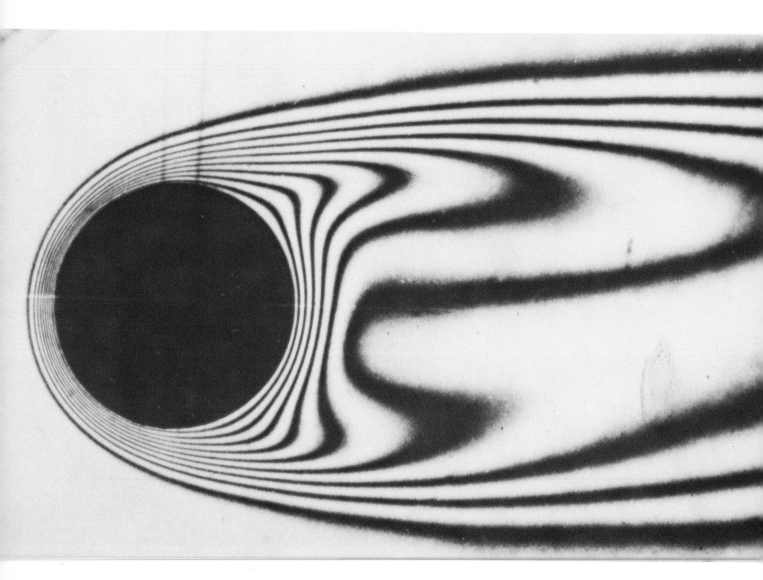

202. Cooling of a cylinder in a stream. A Mach-Zehnder interferometer shows isotherms in the rather thick thermal boundary layer at Reynolds number 120, and the laminar separated wake. A 1-in-diameter heated copper cylinder spans the fully developed laminar flow of air in a rectangular channel. Vertical asymmetry results from free convection. *Eckert & Soehngen 1952*

203. Plane convection plume rising from a heated horizontal wire. A thin wire 6 in. long is heated electrically in atmospheric air. Each fringe in this interferogram represents a temperature difference of 4.4°C. The reference grid wires are spaced ½ by ¼ in. In good accord with self-similar solutions of the boundary-layer equations, the width of the plume grows as the $^2/_5$-power of height. *Gebhart, Pera & Schorr 1970*

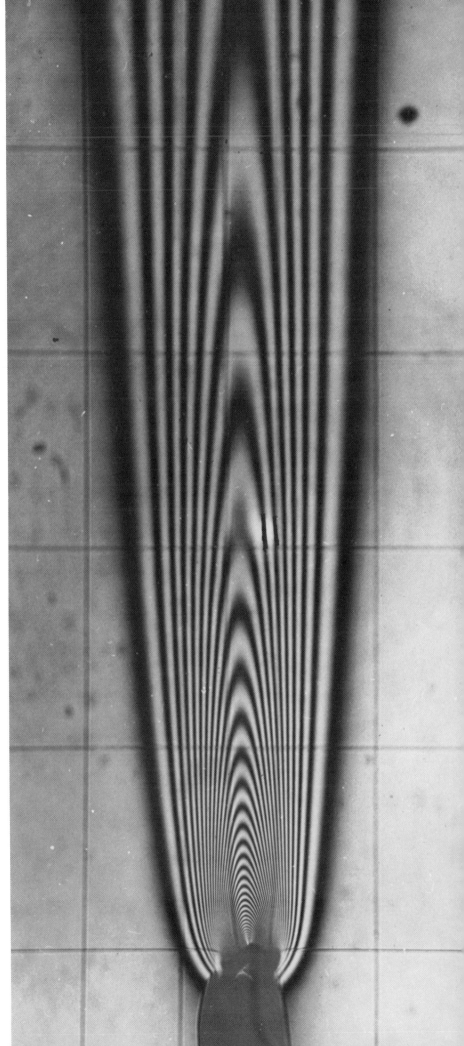

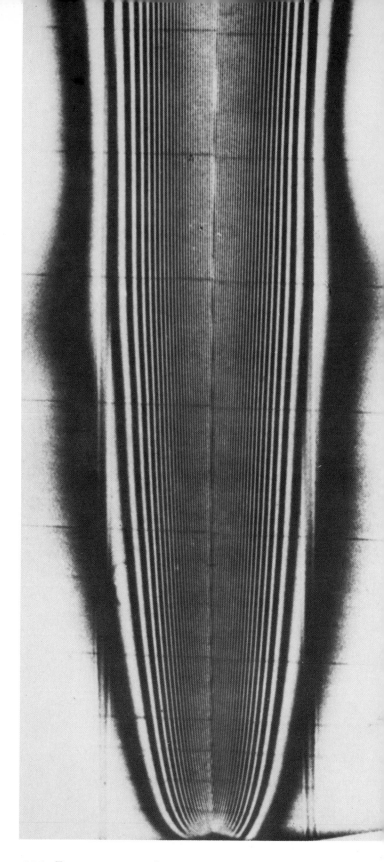

204. Free convection from a vertical plate. The plate is uniformly heated in air, producing a steady laminar flow. An interferogram shows lines of constant density which, at nearly constant pressure, are also isotherms. The Grashof number is approximately five million at a distance of 0.1 m from the lower end of the plate, so that the thermal boundary layer is rather thick. *Eckert & Soehngen 1948*

205. Free convection from a vertical foil. A stretched metal foil is heated electrically in nitrogen at 17 atmospheres. The flux Grashof number is 5×10^{11} at the top, 19 cm from the lower edge. The thin laminar thermal boundary layer has been magnified six-fold in thickness by an anamorphic lens system. For this uniform surface-flux condition the boundary-layer thickness grows as the $\frac{1}{5}$-power of height. *Gebhart 1971*

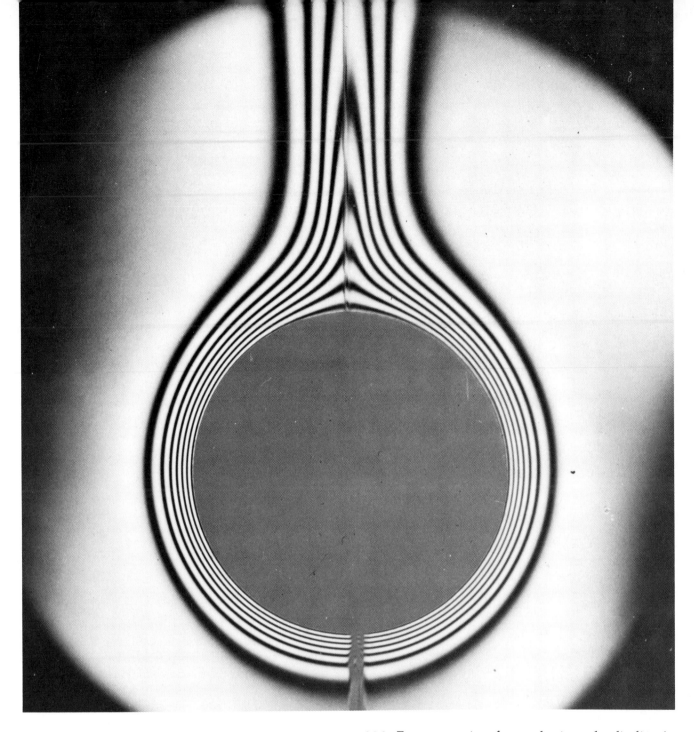

206. Free convection from a horizontal cylinder. A circular cylinder of diameter 6 cm and length 60 cm is heated uniformly 9°C above the ambient air, giving a Grashof number of 30,000. An interferogram shows the thermal boundary layers merging at the top to start a steady laminar plume like that in figure 203. *Photograph by U. Grigull and W. Hauf*

207. Thermally driven boundary layers merging at the top of a cylinder. Illuminated plastic particles in water show the streamline pattern as the laminar free-convection boundary layers from either side of a cylinder meet to form a plume. The flow appears not to separate. *Pera & Gebhart 1972*

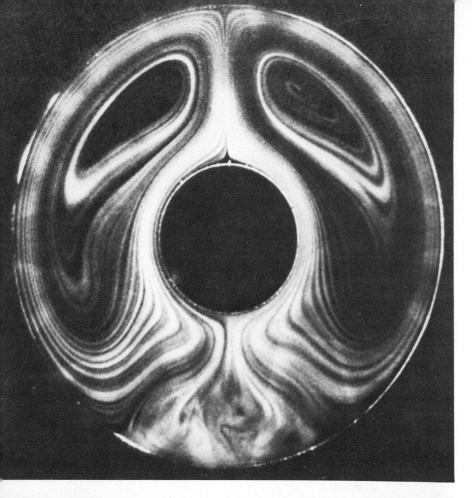

208. Streamlines in convection between concentric cylinders. Cigarette smoke shows the fully developed laminar flow pattern in air at atmospheric pressure. The outer cylinder has three times the diameter of the inner one, and is 14.5°C cooler, giving a Grashof number of 120,000 based on the gap width. *Grigull & Hauf 1966*

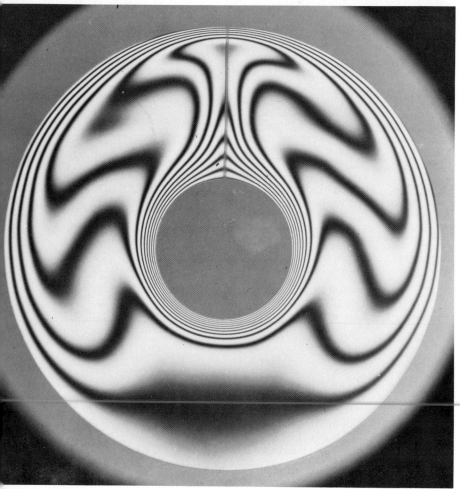

209. Isotherms in convection between concentric cylinders. An interferogram shows the temperature distribution under almost the same conditions as above: ratio of diameters 3.14, temperature difference 13°C, and Grashof number 122,000. Both views show almost stagnant fluid below the inner cylinder. *Grigull & Hauf 1966*

210. Thermal boundary layer beneath a horizontal plate. A laser-holographic interferogram shows isotherms in air, averaged over a 30-cm depth, under the left half of a heated rectangular aluminum plate 10 cm wide and 1 cm thick. The edge of the plate lies directly above the small pattern of concentric dust-particle diffraction rings at lower left. The Grashof number is 3.3 million based on width, and the temperature difference is 2.6°G per fringe. *Hatfield & Edwards 1981*

211. Free convection from three horizontal cylinders. An interferogram shows the laminar convection plume in air from each heated cylinder enveloping the thermal boundary layer of the one above. *Eckert & Soehngen 1948*

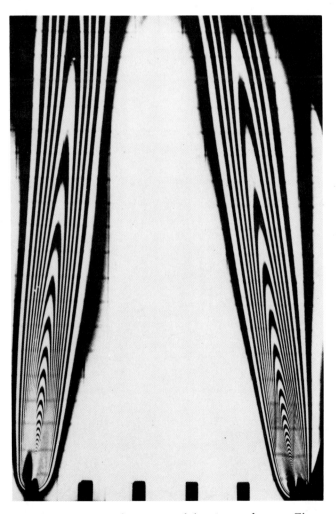

212. Interaction of two equal laminar plumes. Electrically heated nichrome wires 18 cm long and 7.2 cm apart form plumes as in figure 203, which merge by entrainment of the air between. *Pera & Gebhart 1975*

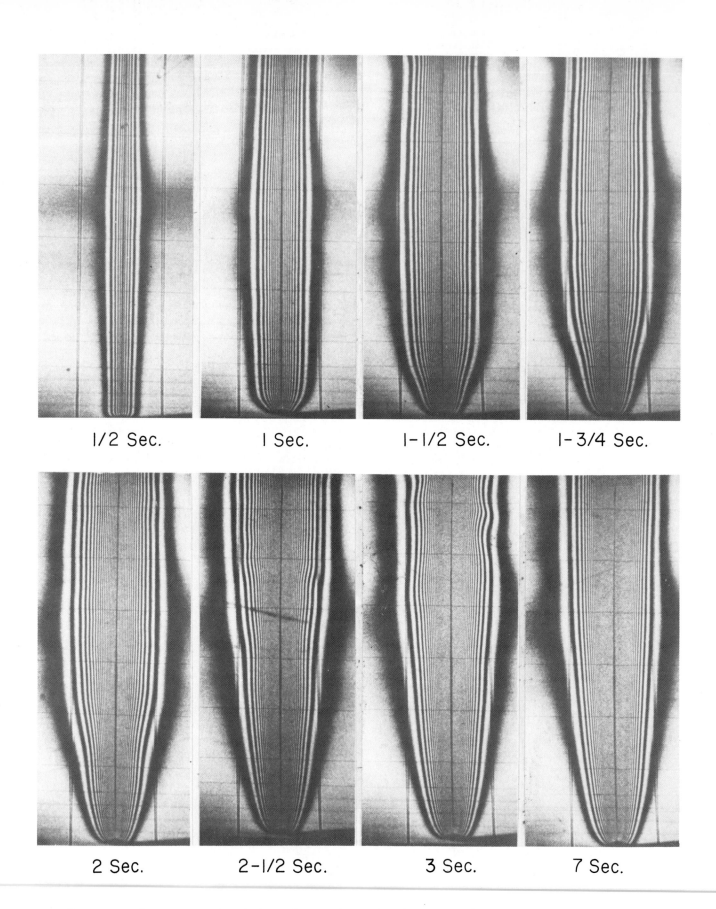

1/2 Sec. 1 Sec. 1-1/2 Sec. 1-3/4 Sec.

2 Sec. 2-1/2 Sec. 3 Sec. 7 Sec.

213. Convection from an impulsively heated vertical foil. A foil 0.013 mm thick is suddenly heated electrically in air at 17 atmospheres. The effect of the leading edge is seen propagating upwards until the steady laminar flow of figure 205 is established. The vertical spacing of the upper grid wires is 2.5 cm. The fringe field is spread optically by a factor of six in the horizontal direction. *Gebhart & Dring 1967*

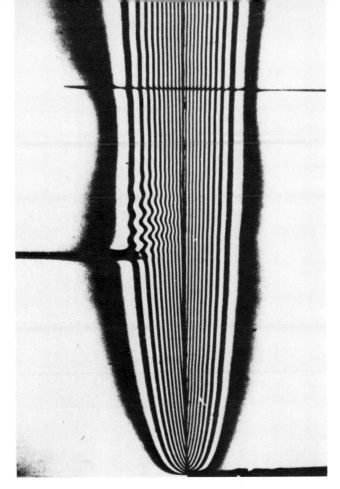 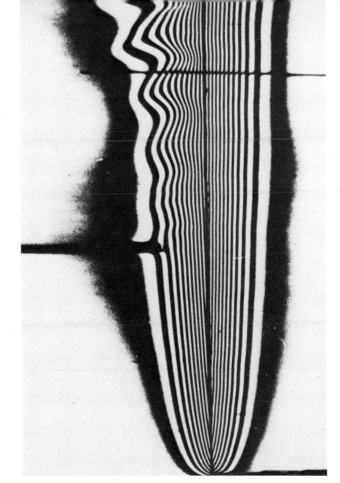

214. Instability in convection from a vertical plate. A ribbon is oscillated in the laminar free-convection boundary layer on an electrically heated foil in nitrogen at 16 atmospheres. Interferograms magnified six-fold in width show disturbances damped on the left at 11.5 Hz but amplified on the right at 3 Hz, in accord with linear stability theory. *Polymeropoulos & Gebhart 1967*

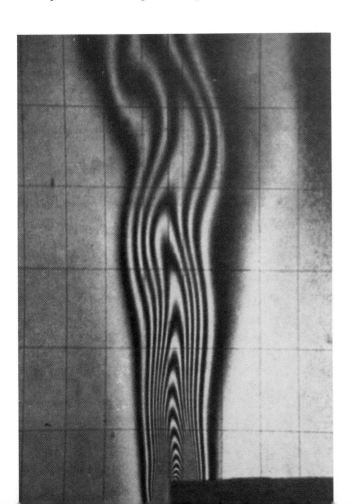

215. Instability of a plane convection plume. A laminar plume rises in atmospheric air from a thin wire extending horizontally 6 inches normal to the photograph, and heated electrically. The interferogram shows temperature differences of 4°C per fringe. The reference grid in the background has ¼- by ½-in rectangles. A metallic ribbon extending the length of the wire is oscillated in the midplane above it at 3.6 Hz. The disturbances are seen to amplify as they are convected upward. *Pera & Gebhart 1971*

216. Steady axisymmetric plume. An interferogram shows the steady laminar convection plume in atmospheric air from a concentrated heat source, a diffusion flame of 0.5 mm diameter. *Pera & Gebhart 1975*

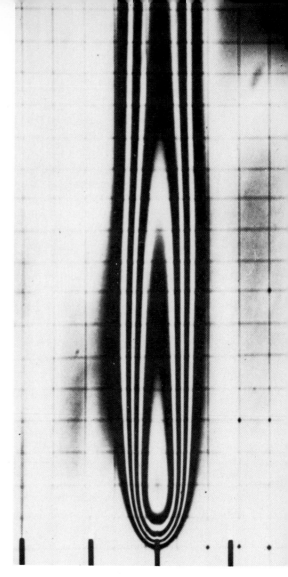

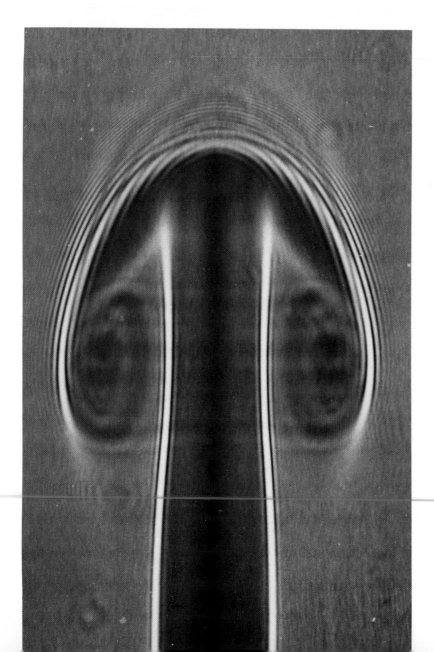

217. Axisymmetric starting plume. A shadowgraph shows the cap of the laminar plume rising from an electrode that has impulsively started injecting heat at 1 cal/s into a concentrated solution of sodium carbonate. The cap rises at a constant speed of 1.2 cm/s and grows by geometric similarity with its width, here 1 cm, equal to a fifth of its height. Below the cap, the axial column is like that in the steady plume shown above. *Shlien & Boxman 1981*

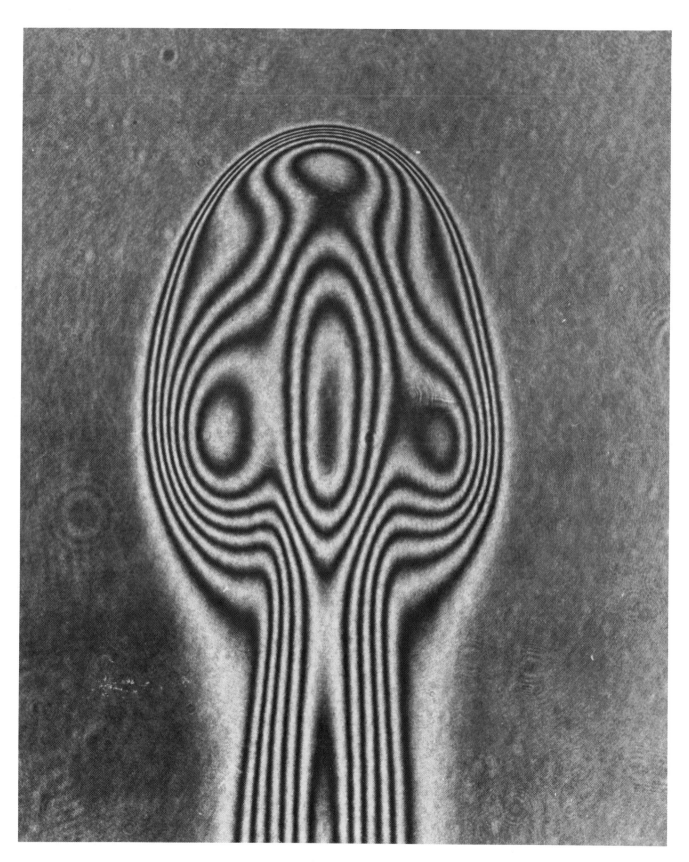

218. Cap of an axisymmetric starting plume. The laminar cap is 1 cm wide after having risen 7 cm from a small electrode at the bottom of a tank of 0.1 per cent sodium chloride solution in water. The spiral vortex-ring structure shown in the preceding photograph for a concentrated solution with Prandtl number 21 is absent here at Prandtl number 7. The interferometer fringes are not isotherms because the flow is axisymmetric, but they show temperature maxima on the axis near the top of the cap and off the axis in the middle. *Shlien & Boxman 1981*

9. Subsonic Flow

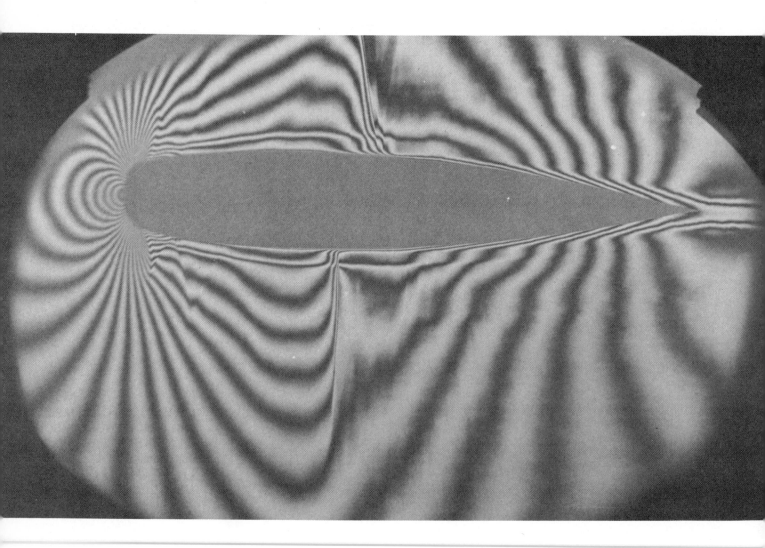

219.   Symmetric airfoil at Mach number 0.8. Transonic flow of air past a profile of 12.7 cm chord and 16.3 per cent thick was photographed in a duct 10 cm wide. The constant-density lines of an interferogram show a flow pattern almost symmetric top and bottom, with shock waves standing near midchord and an attached boundary layer thickening toward the trailing edge. A very small laminar separation region at the shoulder of the profile causes distortion of the expansion fan. *Hiller & Meier 1971*

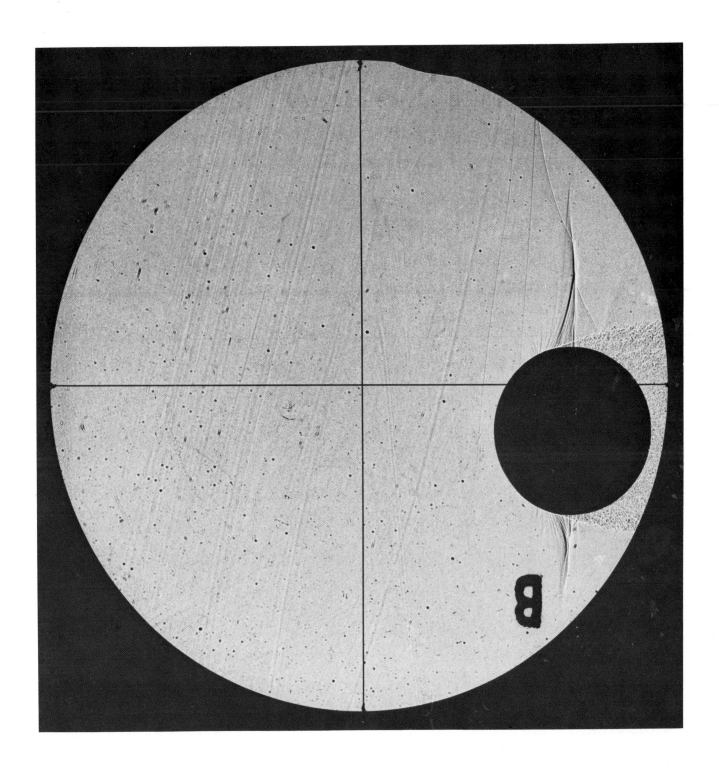

220. Sphere at $M=0.86$**.** A sphere in free flight through air at Reynolds number 920,000 has been caught almost out of the field of view by the flash of a shadowgraph. The shock wave is seen to induce turbulent separation ahead of the equator, in contrast to the situation at low Mach number shown in figure 58. *Photograph by A. Stilp, Ph.D. thesis, Ernst-Mach-Institut, Freiburg, 1965, courtesy of H.-O. Amann*

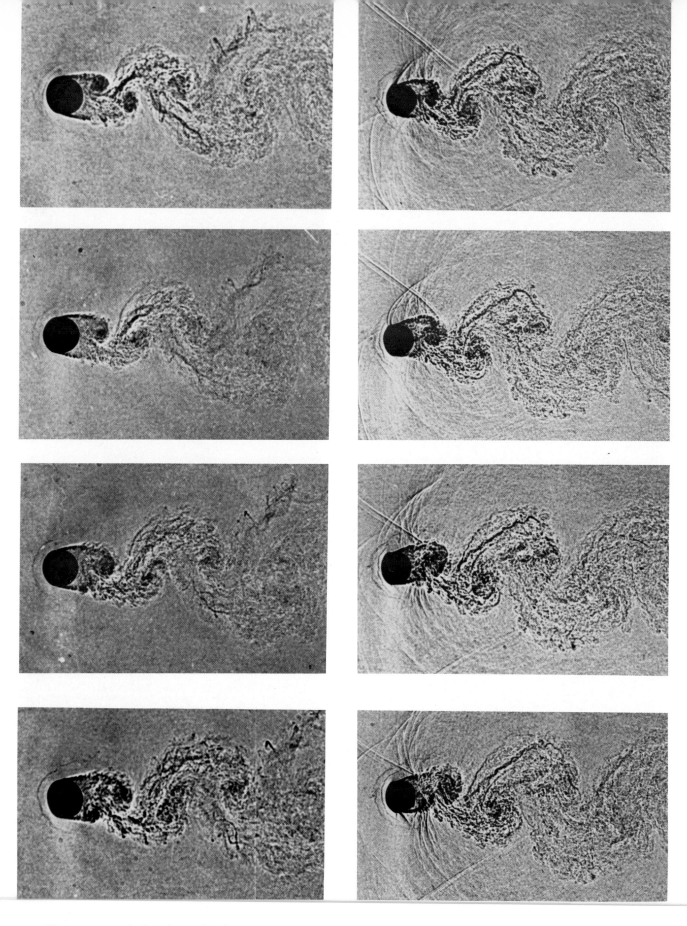

221. Vortex street behind a cylinder. Multiple-spark shadowgraphs taken at the rate of 30,000 per second show one-third of a period of vortex shedding from a circular cylinder. On the left is purely subsonic flow at free-stream Mach number 0.45 and Reynolds number 110,000; on the right is intermittently supercritical flow at M=0.64 and R=1.35 million. The cylinder is slightly distorted by parallax. *Dyment, Gryson & Ducruet 1980*

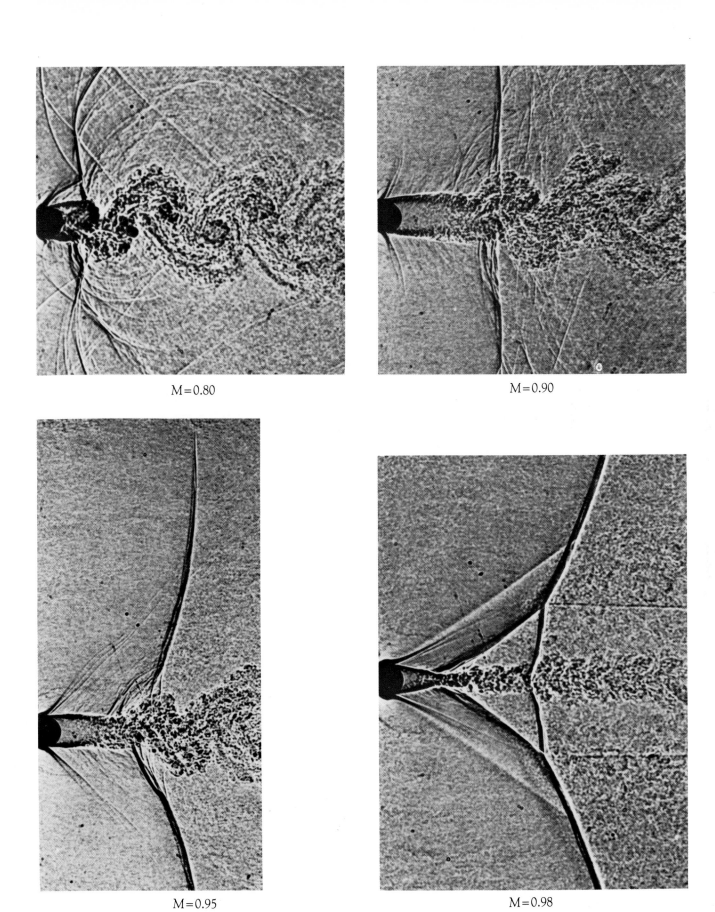

M=0.80

M=0.90

M=0.95

M=0.98

222. Development of the wake with increasing M. As the free-stream Mach number rises toward unity, the wake behind a circular cylinder changes from the periodic shed- ding characteristic of low speeds to the quasi-steady necked wake that persists to supersonic speeds. *Dyment, Gryson & Ducruet 1980*

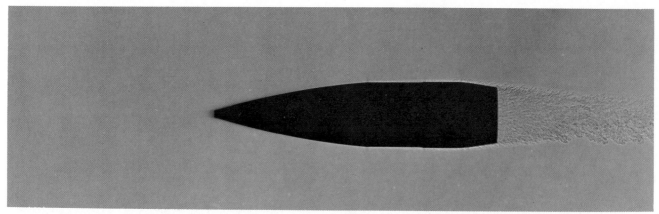

M=0.840

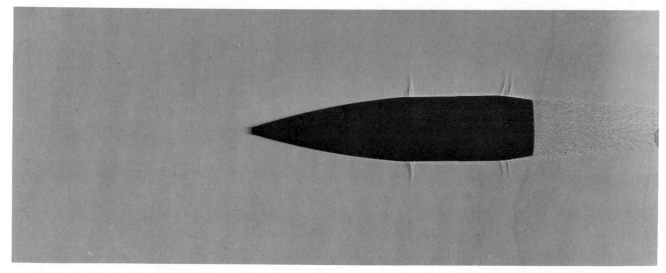

M=0.885

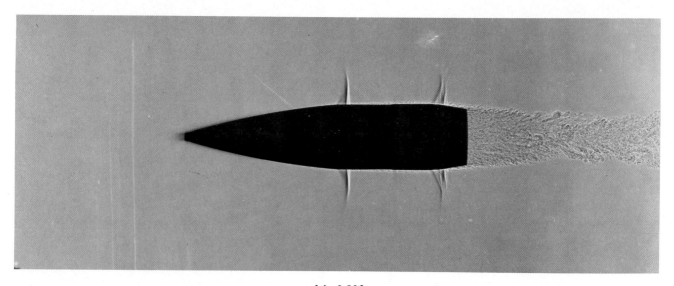

M=0.900

223. Projectile at high subsonic speeds. The spark shadowgraphs on these two pages have been arranged to show the shock-wave pattern growing into the subsonic field around a model of an artillery shell as its Mach number is increased. The shell is in free flight through the atmosphere at less than 1.5° incidence. These five photographs are from four different firings, in each of which the Mach number is gradually decreasing as the shell decelerates. *Photographs by A. C. Charters, in von Kármán 1947*

132

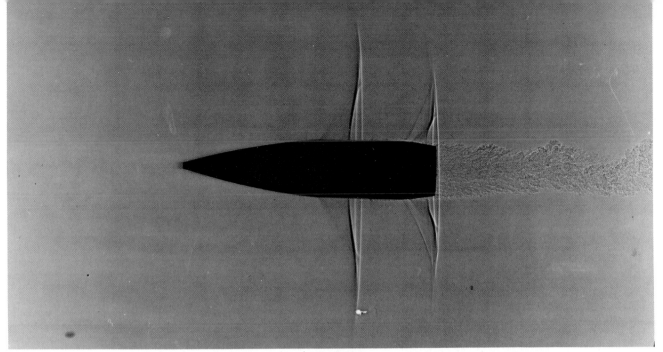

M=0.946

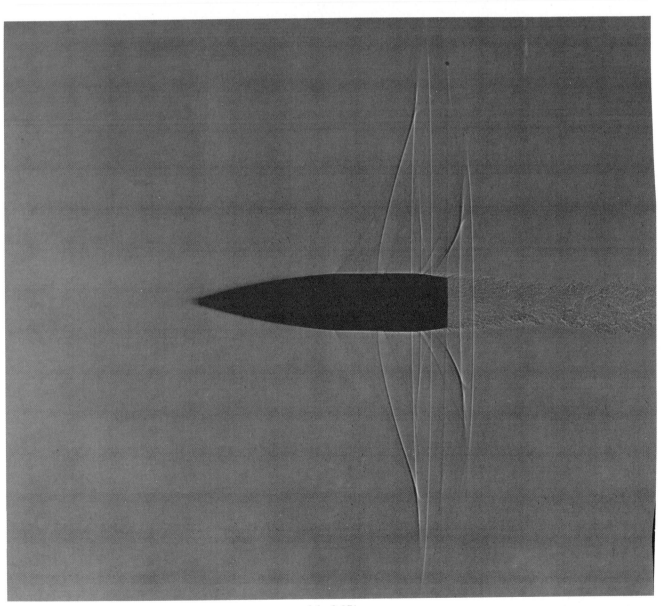

M=0.971

133

M=0.978

224. Projectile at near-sonic speed. Still closer to the speed of sound, the shock-wave pattern of the preceding pages has spread laterally to great distances. These two photographs are from the same firing, so that the second one was actually taken earlier in the trajectory. *Photographs by A. C. Charters, in von Kármán 1947*

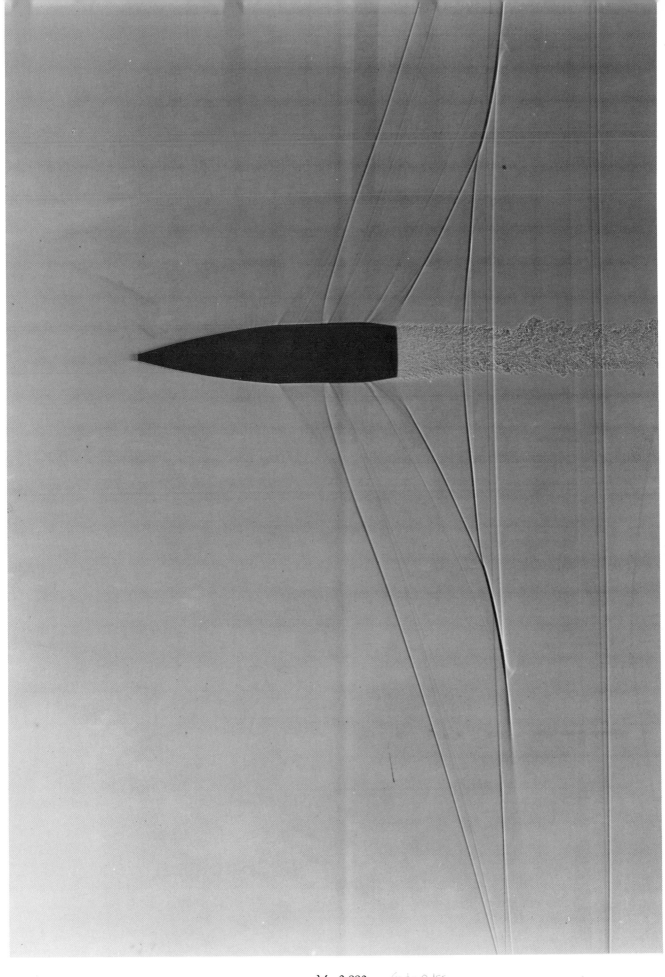

M=0.990   Go to p 155

10. Shock Waves

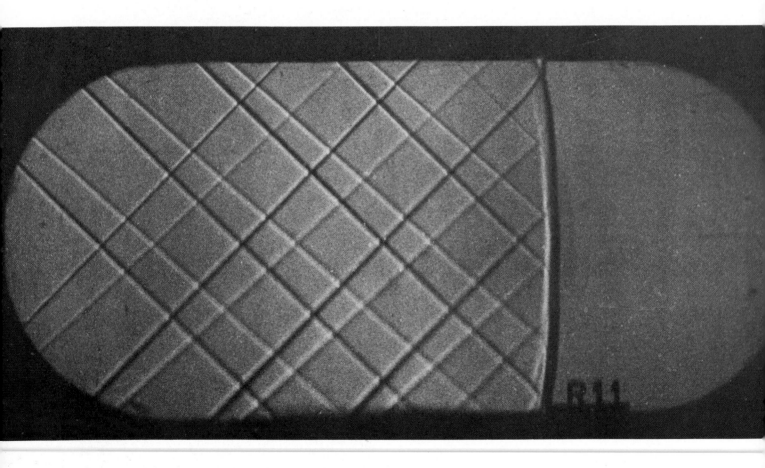

225. Normal shock wave at $M=1.5$. A pattern of pairs of weak oblique shock waves (the N-waves of figures 265 and 269) is produced by strips of tape on the floor and ceiling of a supersonic nozzle. They terminate at an almost straight and normal shock wave, showing that the flow is subsonic downstream. *U. S. Air Force photograph, courtesy of Arnold Engineering Development Center*

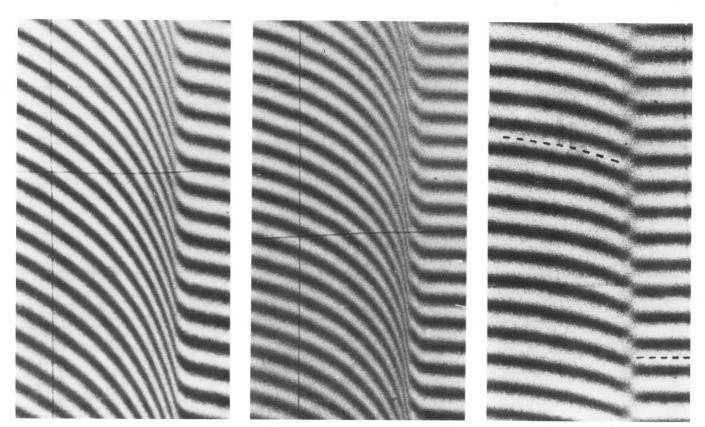

226. Unsteady formation of a normal shock wave.
The classical description of the steepening of a continuous compression into a shock wave is here visualized by interferometry in a shock tube. Because the bursting diaphragm gives a ragged initial wave, a plastic sheet is hung in the shock tube, where the primary wave strikes it head-on. Having finite mass, the sheet accelerates smoothly to form a continuous wave. Three views at successively greater distances from the sheet (out of sight at the left) show the density profiles steepening. They appear discontinuous in the last view, where a dotted line has been drawn to indicate the fringe shift. *Griffith & Bleakney 1954*

227. Steady formation of an oblique shock wave. A cylindrical concave surface in a supersonic wind tunnel at Mach number 1.96 produces a converging fan of compression waves, which are made visible by schlieren photography with the knife edge parallel to the free stream. They focus roughly as a centered compression, forming a strong oblique shock wave that turns the stream through 22.5 degrees. The surface extends upstream as a flat plate at zero incidence so that the weak shock wave from the slightly blunt leading edge will not obscure the view. The surface is roughened to make the boundary layer turbulent, so that it will not separate. *Johannesen 1952*

228. Attached oblique shock waves on a wedge in supersonic flow. Air in a supersonic wind tunnel at M=1.96 is deflected by a sharp wedge of 10° total angle with its lower edge parallel to the stream. A schlieren photograph with horizontal knife edge shows shock waves of pressure ratio 1.7 above and 1.02 below. The laminar boundary layer is visible on the lower surface. *Bardsley & Mair 1951, courtesy of N. Johannesen*

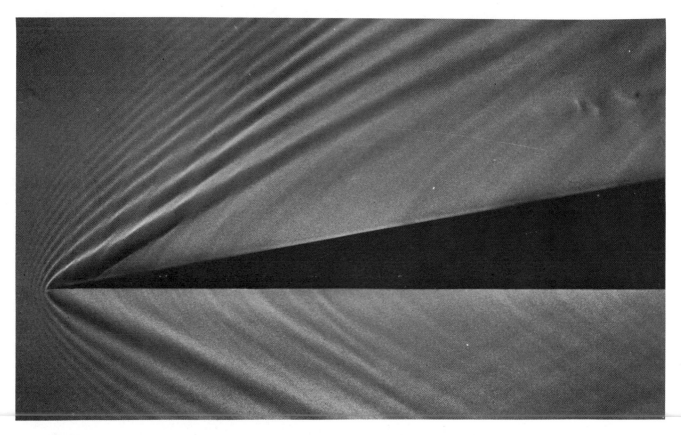

229. Hydraulic analogy for a wedge in supersonic flow. A uniform sheet of water 5 mm deep flows at supercritical speed over a horizontal plane, and is deflected by a sharp 10° wedge with its lower edge parallel to the stream. Gravity waves simulate the oblique waves of the upper photograph, but the pattern is complicated by capillary waves upstream. *Photograph by E. J. Klein, in Merzkirch 1974*

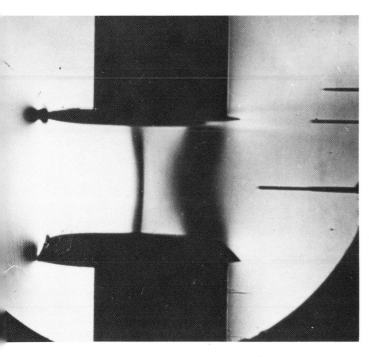

230. Critical and supercritical flow through a straight grid of airfoils. The channel between symmetrical airfoils in a wind tunnel is choked. On the left a weak normal shock wave returns the flow essentially to the same subsonic choking Mach number of 0.65 that is found ahead of the grid. On the right the flow becomes supersonic behind the sonic throat and reaches a maximum Mach number of 1.4 behind the grid, as shown by Mach diamonds from the trailing edges. The schlieren knife edge is horizontal. *Ackeret & Rott 1949*

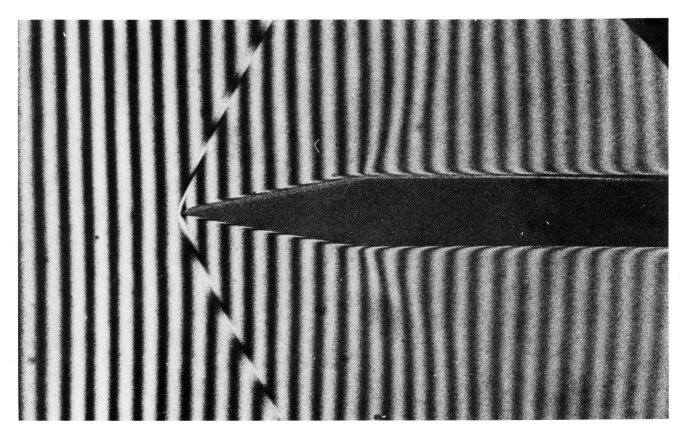

231. Symmetric shock waves on a wedge. A shock tube is here used as a transient wind tunnel. An interferogram shows air flowing at M=1.45 over a wedge-plate of 10° semi-vertex angle. A steady flow has been established 100 microseconds after the incident shock wave passed the tip. *Bleakney, Weimer & Fletcher 1949*

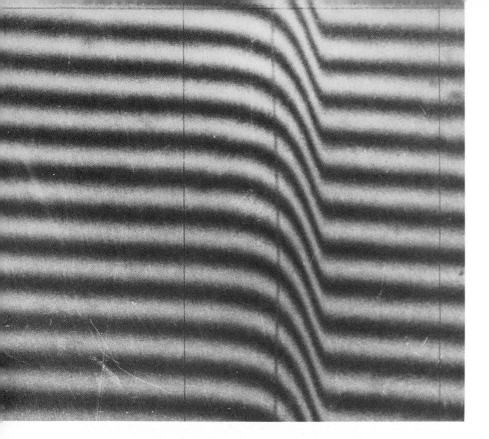

232. Continuous shock-wave structure in carbon dioxide. This interferogram shows a weak plane shock wave propagating to the right through carbon dioxide in a shock tube at a Mach number of about 1.04. Continuous fringes show the fully dispersed structure caused by the relatively slow relaxation of vibrational energy. The thickness of the shock wave is about 8 mm. *Photograph by Walker Bleakney*

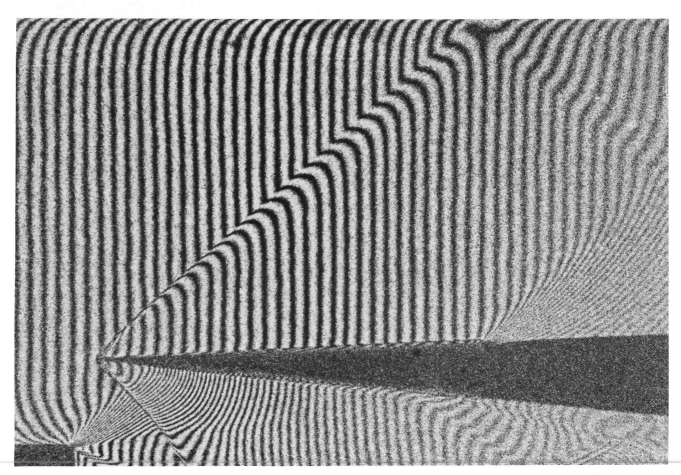

233. Relaxation broadening of the shock wave from a wedge. An interferogram shows steady flow of nitrous oxide in a shock tube at Mach number 1.67. The upper surface of the wedge is inclined 2° to the stream. The shock wave broadens as the vibrational energy, frozen at the tip, relaxes to give a fully dispersed structure. *Hornby & Johannesen 1975*

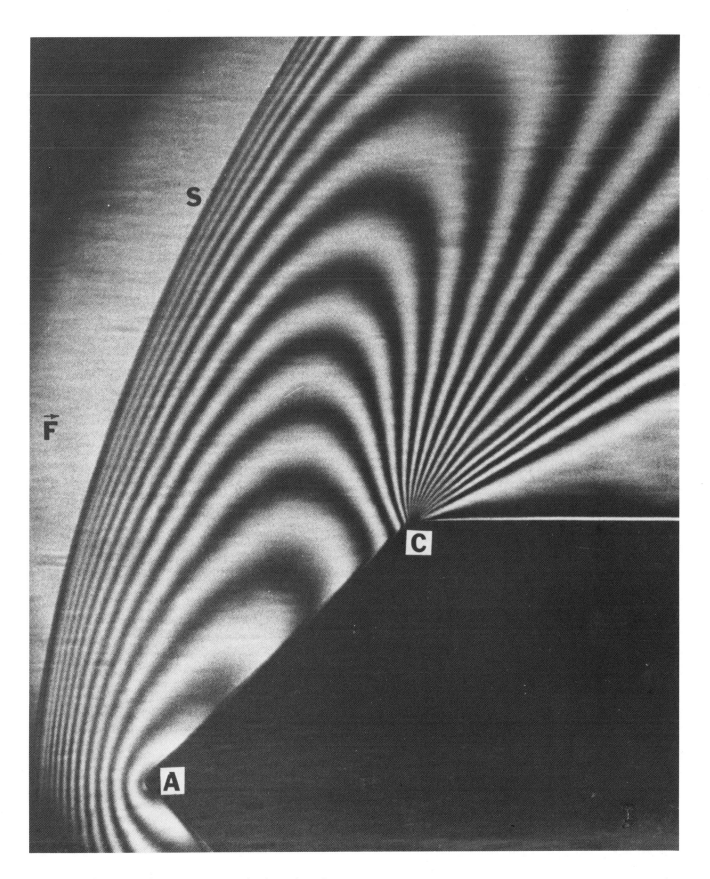

234. Detached bow shock wave on a thick wedge. An infinite-fringe interferogram shows lines of constant density about a wedge A of 45° half angle in a stream of air F at Mach number 2.5. No attached shock wave can turn the flow, so the shock wave S detaches to produce a zone of subsonic flow extending back to the corner C. There a Prandtl-Meyer expansion fan turns the flow to a supersonic region behind the corner. *Glass 1974*

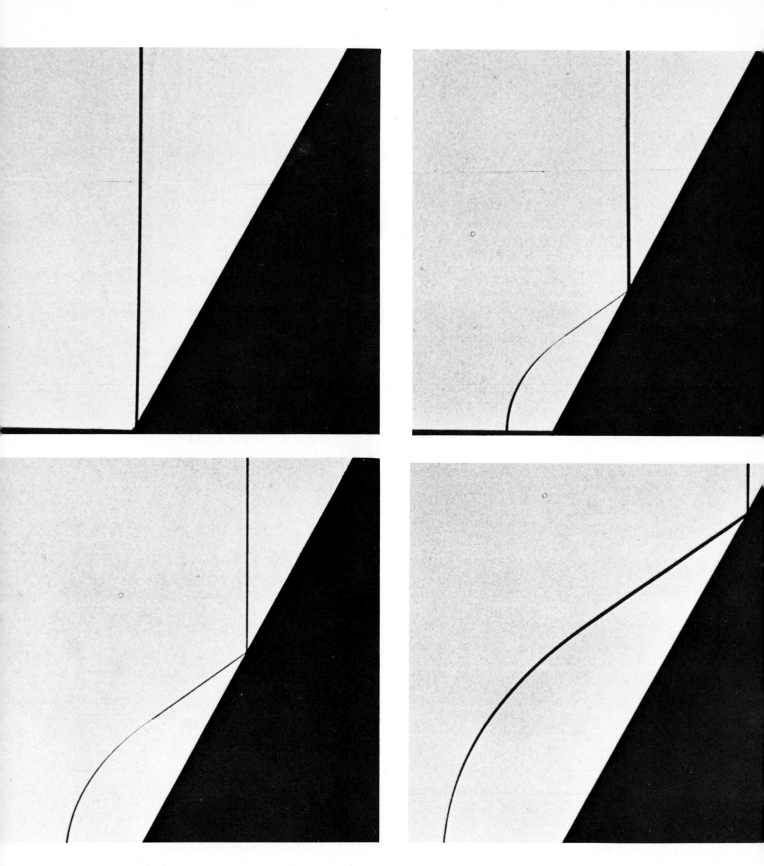

235. Regular reflection of a shock wave from a wedge.
Shadowgraphs show a vertical plane shock wave striking a
60° wedge from the left. The wedge angle and shock-wave
strength are such that the reflection is regular. The angles
of the incident and reflected shock waves to the face of the
wedge are not equal, because the phenomenon is nonlinear. Insofar as viscosity is negligible, there is no reference length available, so the flow is self-similar: the wave
pattern is seen to expand linearly in time from the instant
of contact with the leading edge. *Schardin 1965, courtesy of
H. Oertel, Sr.*

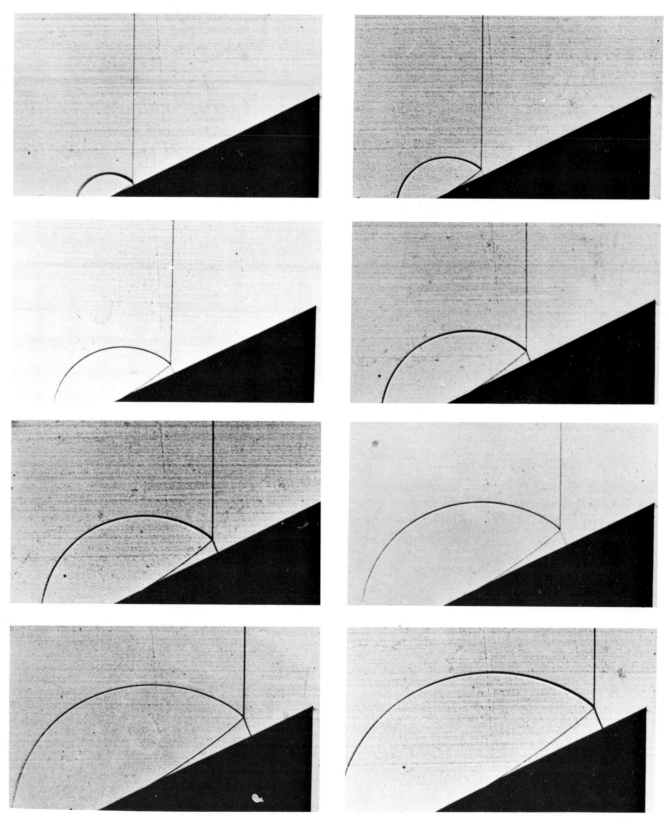

236. Mach reflection of a shock wave from a wedge.
Shadowgraphs show a vertical plane shock wave striking a 25° wedge. The regular reflection shown on the opposite page is replaced for smaller wedge angles by this pattern of Mach reflection. A third shock wave, the Mach stem, runs normal to the surface and joins the incident and reflected shock waves at the triple point. The curve running down left from that point to meet the surface smoothly is the slip line, across which the entropy is discontinuous because the air on each side has been subjected to different shock waves. The time between photographs is 6×10^{-5} s. *Schardin 1965, courtesy of H. Oertel, Sr.*

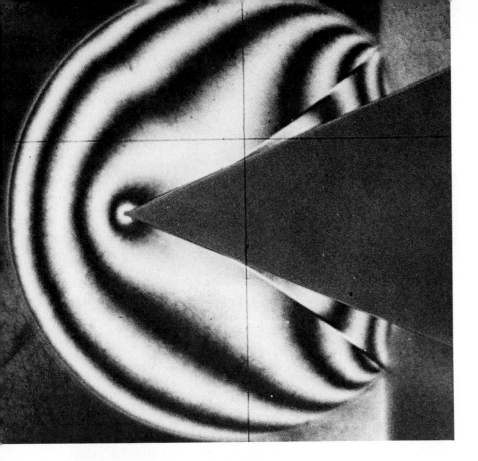

237. Ordinary Mach reflection from a wedge. This infinite-fringe interferogram shows lines of constant density behind a plane shock wave impinging symmetrically on a wedge of 22.5° half angle. The density discontinuity is evident across each slip line. *Griffith & Bleakney 1954*

238. Irregular Mach reflection from a wedge. A shadowgraph shows that when the incident shock wave is strong—here moving at 2400 m/s—ordinary Mach reflection is replaced by so-called irregular Mach reflection. Here the Mach stem of the original Mach reflection has just reached the base of the wedge. The new feature is that the reflected shock wave above the wedge consists of a straight and a curved section, and from their juncture another shock wave extends perpendicular to the slip line from the primary triple point. Near the wedge the slip line curls due to boundary-layer effects. *Prasse 1977, courtesy of H.-O. Amann*

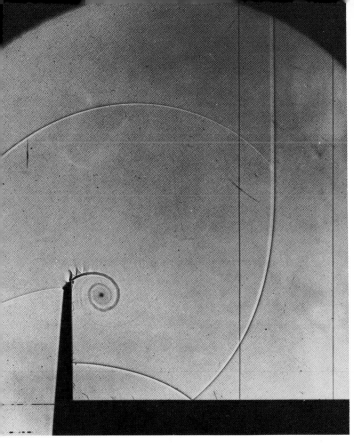

239. Diffraction of a shock wave over an edge. These two successive shadowgraphs show how the vortices in figures 82 and 83 are generated in a shock tube. A relatively weak plane shock wave passes over a vertical edge, producing a slip line that rolls up. Here a series of lambda shock waves stand on the slip line. In the second photograph the reflected shock wave has intersected and been kinked by the vortex sheet, in which an instability is beginning to be evident. A weak shock wave can be traced from the lambda shocks through the vortex to an intersection at right angles with the main reflected wave. This is the segment of the reflected shock wave that has been swept through the left-hand side of the vortex. *Photographs taken by Russell E. Duff in Otto Laporte's laboratory at the University of Michigan in 1948-49*

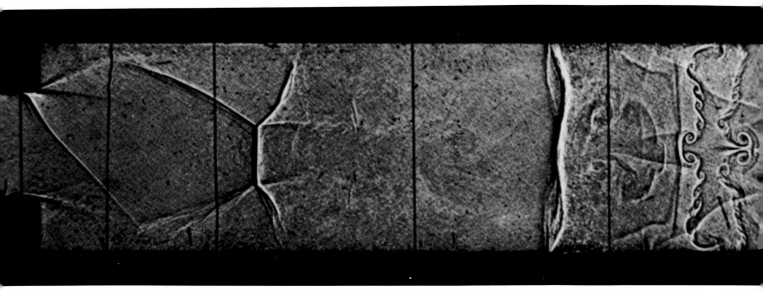

240. Diffraction of a shock wave inside a box. A shock wave in nitrogen is diffracted through a window at one end of a rectangular box and reflected from the other end. A shadowgraph shows a remarkable pattern of shock waves, slip lines, and vortices, but one that is altogether determinate and reproducible. The three rope-like traces at the right are slip lines generated as the diffracted shock wave oscillated in shape moving across the box, which have been perturbed by shock waves passing over them roughly at right angles. Several examples of separated boundary-layer flow are also evident. *Photograph by Russell E. Duff in Laporte's laboratory*

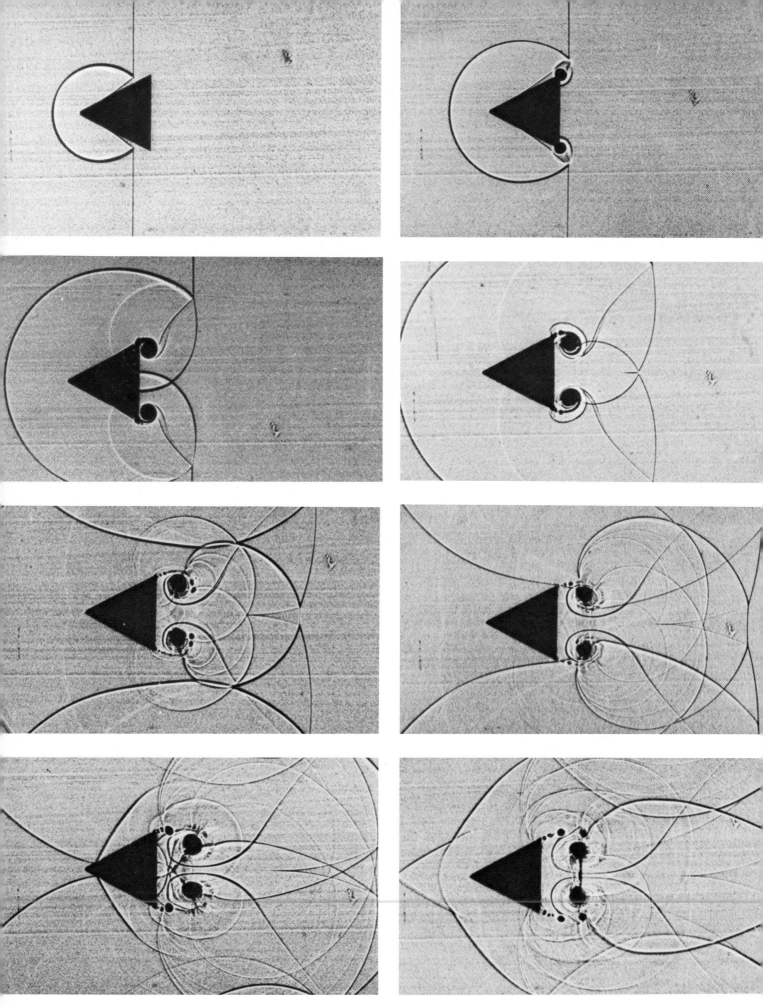

146

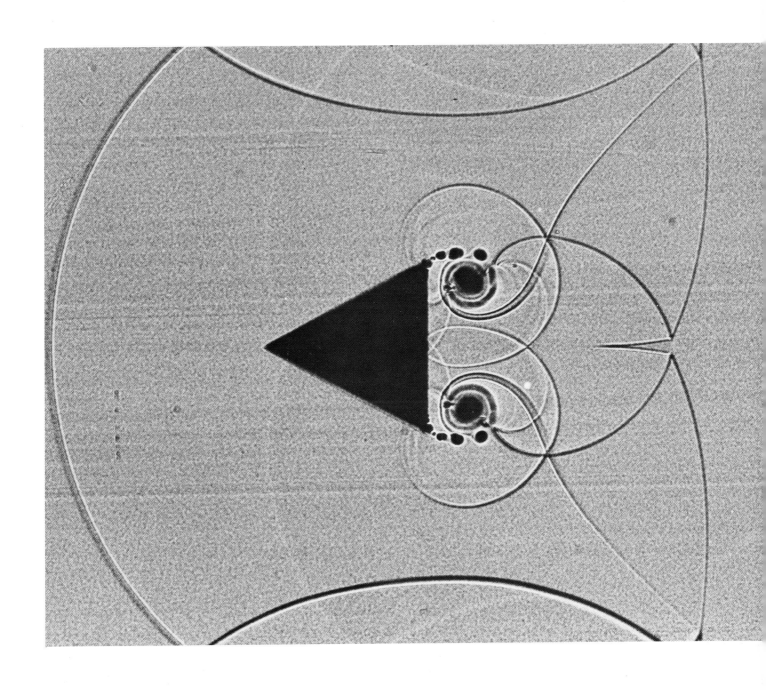

241. Diffraction of a shock wave by a finite wedge.
The first shadowgraph of the sequence on the opposite page shows ordinary Mach reflection of a plane shock wave, as in figures 236 and 237. As the shock wave passes the base, the flow separates and vortex sheets are generated that roll up as in figure 81. Further interaction produces an increasingly elaborate pattern of shock waves, slip lines, and vortices. The enlargement above shows the remarkably complicated symmetric flow at an instant between the fourth and fifth photographs on the opposite page. *Photographs by H. Schardin, in Oertel 1966*

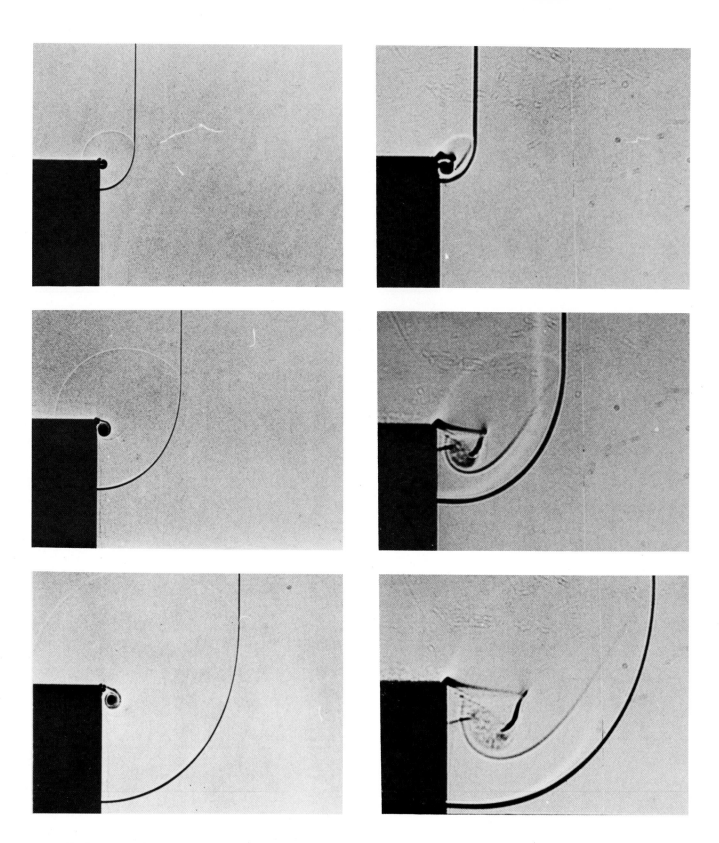

242. Diffraction of a weak shock wave down a step.
The shadowgraphs above show a plane shock wave of
Mach number 1.3 in a shock tube. The pattern of vertical
incident shock wave, curved diffracted wave, and circular
expansion wave grows linearly in time. Viscous separation
at the corner generates a rolled-up vortex sheet. *Schardin
1965, courtesy of H. Oertel, Sr.*

**243. Diffraction of a stronger shock wave down a
step.** With the Mach number of the incident shock wave
increased to 2.4, the flow pattern still grows linearly in
time, but shows a more complicated structure than in the
sequence at the left. The flow is supersonic at the corner,
so that no disturbances propagate upstream. *Schardin 1965,
courtesy of H. Oertel, Sr.*

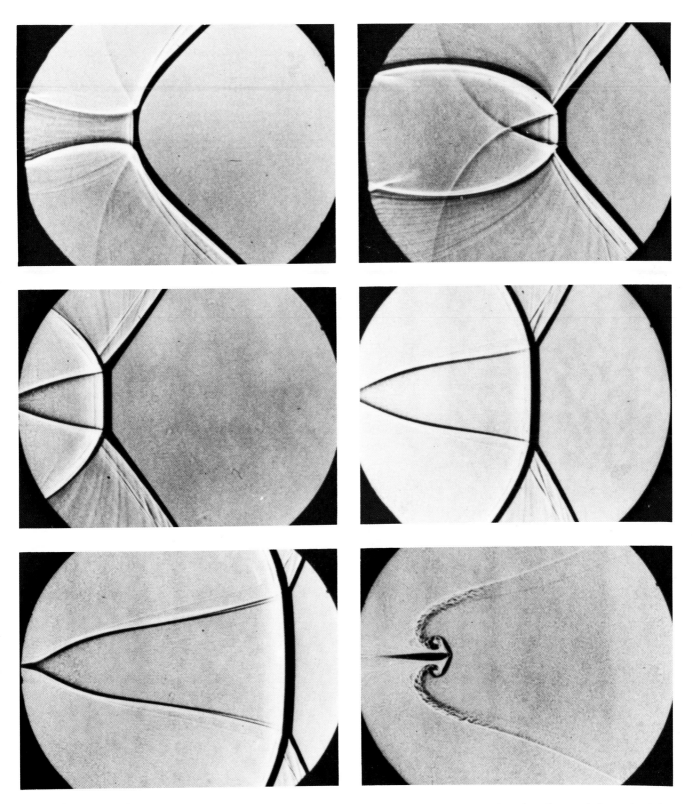

244. Focusing of a shock wave. A plane wave of shock Mach number 1.3 coming from the right has reflected from a parabolic cylinder that is just visible at the left edge of the window in the top two shadowgraphs. In the first picture the pattern consists of a central circular arc of reflected shock wave joined, by almost horizontal expansion waves, to shock waves diffracted from the corners of the parabolic reflector. In the second picture the expansion waves have crossed each other and formed triple-shock intersections by Mach reflection. In the remaining photographs the reflector has been shifted one window radius to the left. The almost plane Mach stem leaves high-entropy air at the focal spot. In the last view the shock waves have passed, and the slipstreams from the Mach reflection become turbulent, pinch off, and roll up due to mutual interaction. *Sturtevant & Kulkarny 1976*

245. Multiple shock waves on an airfoil with a laminar boundary layer. A zone of local supersonic flow ordinarily terminates in shock waves, whose form is influenced by the boundary layer. When it is laminar, a succession of so-called lambda shock waves first appears. This schlieren photograph with vertical knife edge shows a biconvex airfoil 12 per cent thick at free-stream Mach number 0.8. *Photograph by H. W. Liepmann*

246. Single lambda shock wave on an airfoil with a laminar boundary layer. As the Mach number is increased the multiple shock waves of the upper photograph merge into one. Here a horizontal knife edge shows that at M=0.9 the laminar boundary layer has separated from the biconvex airfoil ahead of the shock wave and become turbulent. *Photograph by H. W. Liepmann*

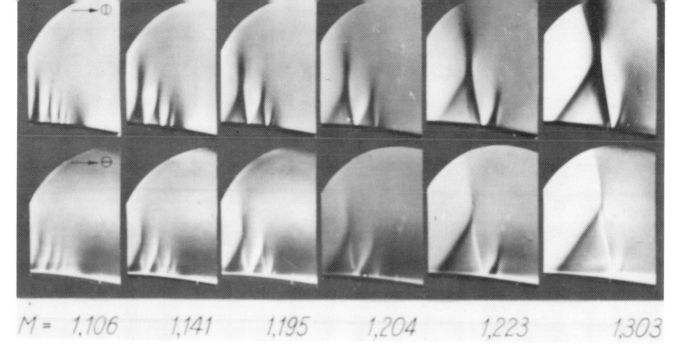

$$M = 1.106 \qquad 1.141 \qquad 1.195 \qquad 1.204 \qquad 1.223 \qquad 1.303$$

247. Shock waves on a laminar boundary layer with increasing Mach number. The lambda shock waves are seen to merge with increase of the local Mach number, whose value is given below each pair of schlieren photographs. Here the laminar boundary layer forms on a curved plate that produces a growing limited region of supersonic flow embedded in a subsonic flow, imitating supercritical flow over an airfoil. The symbols at the left indicate that the knife edge is vertical in the upper row and horizontal below. *Ackeret, Feldmann & Rott 1946*

248. Shock wave on an airfoil with a turbulent boundary layer. When the boundary layer is turbulent, a single shock wave appears. The flow then resembles the inviscid model more closely than when the layer is laminar. Here the free-stream Mach number is 0.84, and the knife edge is vertical to show the shock wave clearly. *Photograph by H. W. Liepmann*

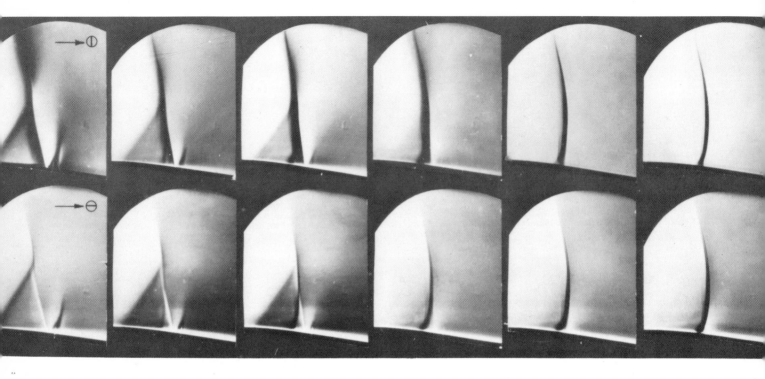

249. Shock waves on a laminar boundary layer becoming turbulent. The local Mach number on a curved plate remains almost fixed at 1.2 or 1.3 as the Reynolds number is doubled, progressing from 1,320,000 at the left to 2,680,000 at the right. As the boundary layer changes from laminar to turbulent ahead of the shock wave, the oblique leg of the lambda shock wave gradually disappears. *Ackeret, Feldmann & Rott 1946*

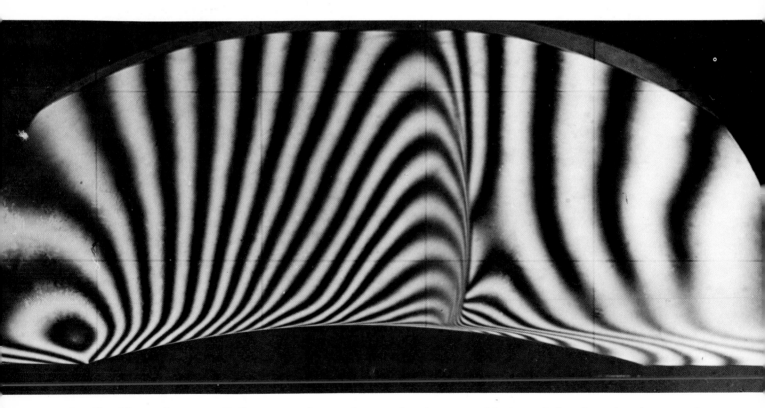

250. Shock wave in transonic flow over a bump. An infinite-fringe interferogram shows transonic flow over a 7-per-cent-thick circular-arc bump on a channel wall. The local region of supersonic flow terminates in a shock wave that interacts with the turbulent boundary layer on the wall, as in the preceding two photographs. *Délery, Chattot & Le Balleur 1975*

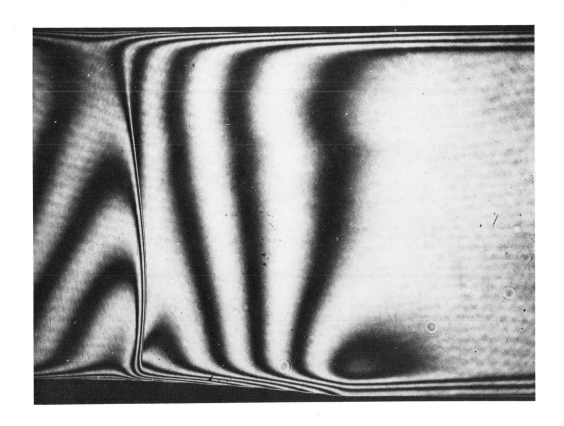

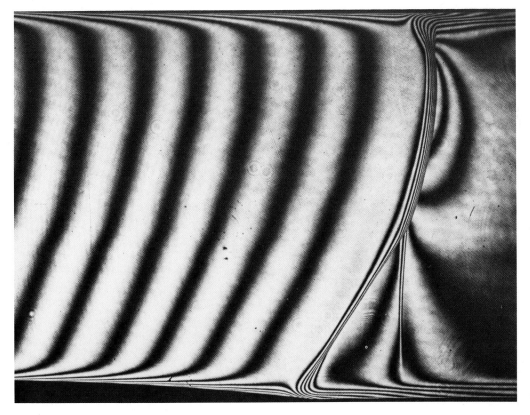

251. Shock wave interacting with a turbulent boundary layer in a channel. Infinite-fringe holographic interferograms show transonic flow over a long bump with a circular-arc rear. In the upper photograph a weak normal shock wave, with maximum Mach number 1.1 upstream, stands on the profile. In the lower photograph a lambda shock wave springs from the trailing edge. *Délery, Surget & Lacharme 1977*

11. Supersonic Flow

252. Axisymmetric model separating from its sabot. A finned body of revolution has been fired from a gun against the supersonic stream of a wind tunnel to give a net free-stream Mach number of 2. A shadowgraph shows the flow tearing away the fingers of the plastic sabot that protected the model during firing. It has broken a grid of fine wires to trigger the spark, just as in Mach's schlieren photograph of 1888 shown in the frontispiece. *NASA photograph, courtesy of W. G. Vincenti*

154

253. Projectile at M=1.015. The model artillery shell of figures 223 and 224 is shown here still earlier in its trajectory, when it is flying at a slightly supersonic speed. A detached bow wave precedes it, and the distant field is quite different, but the pattern near the body is almost identical to that shown in figure 224 for a slightly subsonic speed. This illustrates how the near field is "frozen" as the free-stream Mach number passes through unity. *Photograph by A. C. Charters*

254. Rifle bullet at M=1.1. Two different optical techniques show the wave pattern and wake of a rifle bullet in slightly supersonic flight through atmospheric air. The shadowgraph in the upper photograph is sensitive to changes in the second derivative of gas density. The schlieren photograph below shows gradients of density normal to the knife edge, which is vertical. (The frontispiece shows an earlier version.) The large arcs are intersections of the axisymmetric shock waves with the glass windows of the transonic range. *Photographs by P. Wegener*

255. Airplane model in free flight at M=1.1. Shadowgraphs show a winged model launched into atmospheric air from a gun, as in figure 252. The bow wave is marginally attached at this slightly supersonic speed. In the plan view above, the wings are lifting, as shown by trailing vortices from the tips. In the side view below, the herringbone pattern is produced by pressure pulses from grooves in the wing that trip the boundary layer to make it turbulent over the rear half. *NASA photographs, courtesy of W. G. Vincenti*

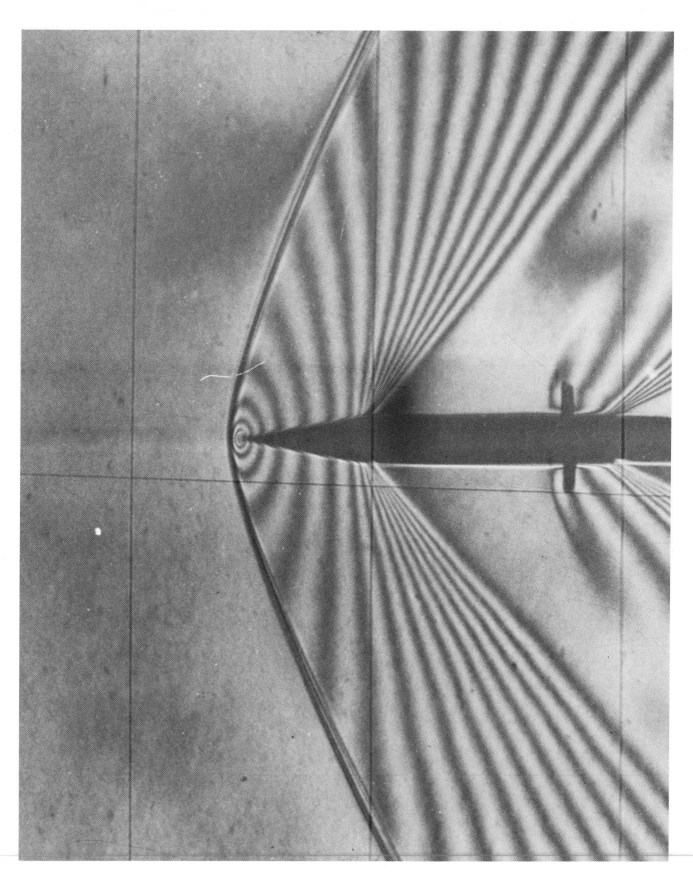

256. Detached bow wave on a thin wedge. An infinite-fringe interferogram shows lines of constant density for a wedge-plate of 10° semi-vertex angle in a supersonic wind tunnel at free-stream Mach number 1.32. The bow shock wave cannot attach to this wedge in air, to give the conical field of figure 228, until the Mach number exceeds 1.4. *Ashkenas & Bryson 1951, courtesy of H. Ashkenas*

257. Attached bow wave on a 22.5° cone at M=1.96. Here the flow is supersonic everywhere between the bow shock wave and the surface of the cone. Hence the field is conical and the wave straight until it is intersected by the expansion wave from the base of the cone. The schlieren knife edge is horizontal, making the image antisymmetric from top to bottom. *Photograph by A. W. Sharp, courtesy of N. Johannesen*

258. Detached bow wave on a 60° cone at M=1.96. At this Mach number in air the bow shock wave cannot remain attached to the vertex of a cone above a semi-vertex angle of 40°. It detaches, forming an embedded subsonic zone extending back to the cone surface. The two crinkled curves downstream are the intersections of the bow wave with the turbulent boundary layers on the wind-tunnel windows. *Photograph by A. W. Sharp, courtesy of N. Johannesen*

259. Cone-cylinder in supersonic free flight. A cone-cylinder of 12.5° semi-vertex angle is shot through air at M=1.84. The boundary layer becomes turbulent shortly behind the vertex, and generates Mach waves that are visible in this shadowgraph. *Photograph by A. C. Charters*

260. Shock-free forebody. At Mach number 2.1 the concave nose of this body of revolution shows the smooth compression required for a shock-free diffuser. The bow shock wave forms only away from the body, as the axisymmetric counterpart of figure 227. The tip was unfortunately bent during firing, which generated the weak shock wave there. *Photograph from Transonic Range, U.S. Army Ballistic Research Laboratory*

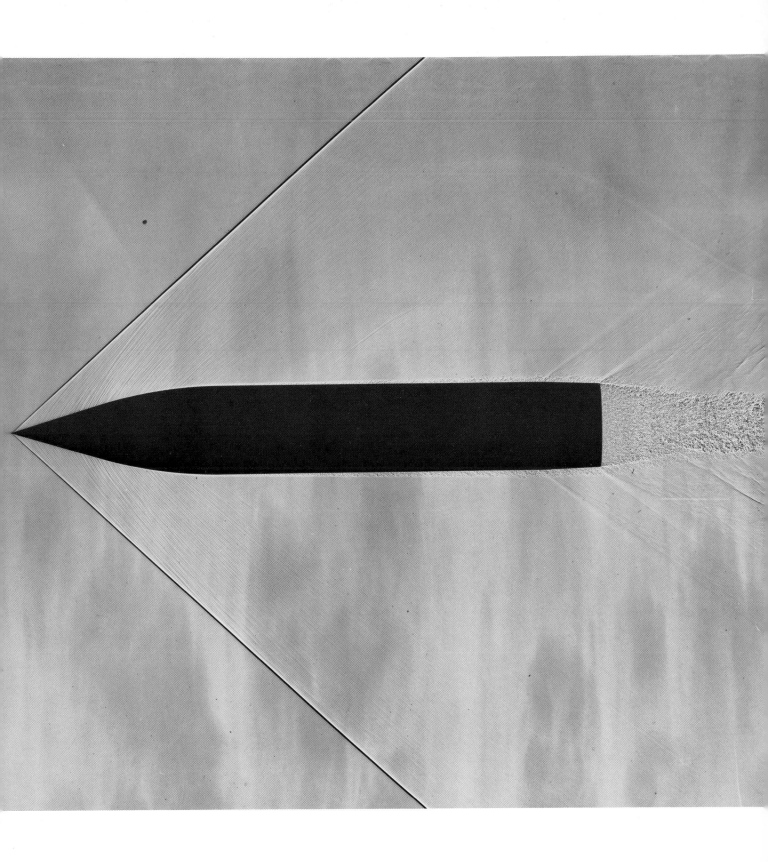

261. Ogive-cylinder in free flight at M=2.58. This shadowgraph shows in remarkable detail the supersonic flow past a long body of revolution at small incidence: an initially straight bow wave from a finite conical tip, weak shock waves generated by roughness on the nose, a grow- ing turbulent boundary layer over the cylindrical midsection, expansion fans at the start of the conical boattail and at the base, and transient shock waves from an unsteady turbulent wake. *Photograph from Transonic Range, U.S. Army Ballistic Research Laboratory*

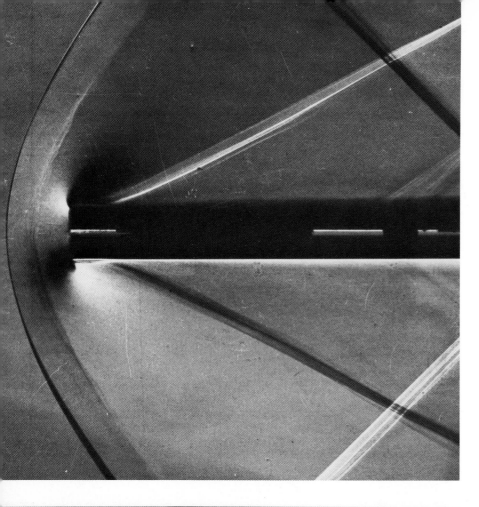

262. Flat-faced plate at $M=2.0$. The knife edge is horizontal in this schlieren photograph, so that light above is dark below. Gaps in the middle of the plate do not affect the air flow. The curved line just downstream of the detached bow shock wave is caused by its intersection with the boundary layers on the glass sides of the wind tunnel. The boundary layer separates at the corner, and its reattachment is marked by the formation of oblique shock waves. *Mair 1952, courtesy of N. Johannesen*

263. Flat-faced cylinder at $M=2.0$. Comparison with the photograph above illustrates that the stand-off distance of the bow shock wave is roughly half as great for an axisymmetric body as for a plane body of the same cross section at the same Mach number. The flow pattern is otherwise much the same, showing separation and an oblique shock wave at reattachment. The bow wave intersects the wall boundary layers in two curves at the right. *Johannesen 1958*

264. Cylinder at M = 2.77 in carbon dioxide. A finite-fringe interferogram of a circular cylinder in free flight shows the bow wave wrapped more closely about the flat face than in the preceding photograph, both because the free-stream Mach number is greater and because the adiabatic exponent is lower, being 4/3 for carbon dioxide compared with 7/5 for air. The shock wave at reattachment is visible, followed by a second oblique shock wave from a bump on the cylinder. *Photograph from Air Flow Branch, U.S. Army Ballistic Research Laboratory*

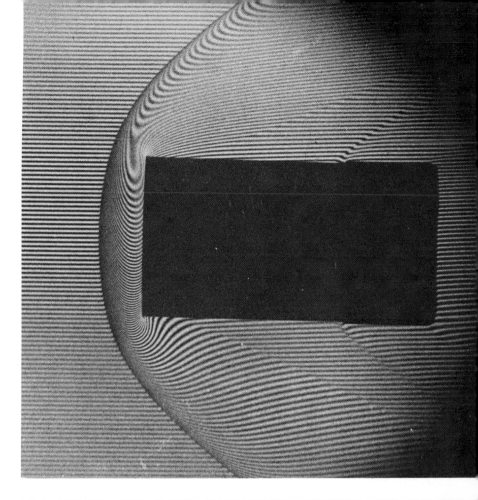

265. Cylinder at M = 3.6 in air. A shadowgraph shows a circular cylinder in free flight at a small negative angle of attack. The apparent squashing of the face is actually optical distortion. The oblique shock wave from boundary-layer reattachment is seen merging with the wave from the wake. At great distances they form the rear of the N-wave pressure signature, shown for a sphere in figure 269, that is characteristic of any object in supersonic flight. *Photograph by A. C. Charters*

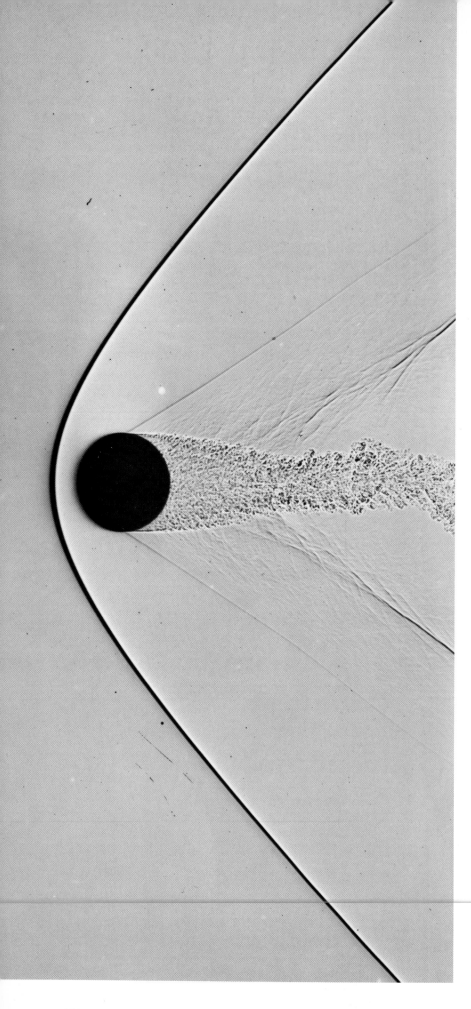

266. Sphere at M=1.53. A shadowgraph catches a ½-inch sphere in free flight through air. The flow is subsonic behind the part of the bow wave that is ahead of the sphere, and over its surface back to 45°. At about 90° the laminar boundary layer separates through an oblique shock wave, and quickly becomes turbulent. The fluctuating wake generates a system of weak disturbances that merge into the second shock wave. *Photograph by A. C. Charters*

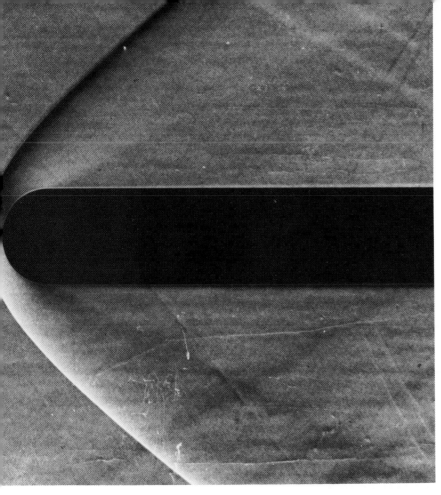

267. Hemisphere-cylinder at M=1.96.
The schlieren knife edge is horizontal to
show the boundary layer on the cylinder.
The Reynolds number is 82,000 based on
nose radius. There is no evidence of separation at the sphere-cylinder juncture. *Mair
1952, courtesy of N. Johannesen*

268. Sphere at M=5.7. A finite-fringe
interferogram displays the detached bow
wave with beautiful clarity, but does not
reveal the details of the boundary layer and
wake that are seen in the shadowgraph on
the next page. *Photograph from U.S. Army
Ballistic Research Laboratory*

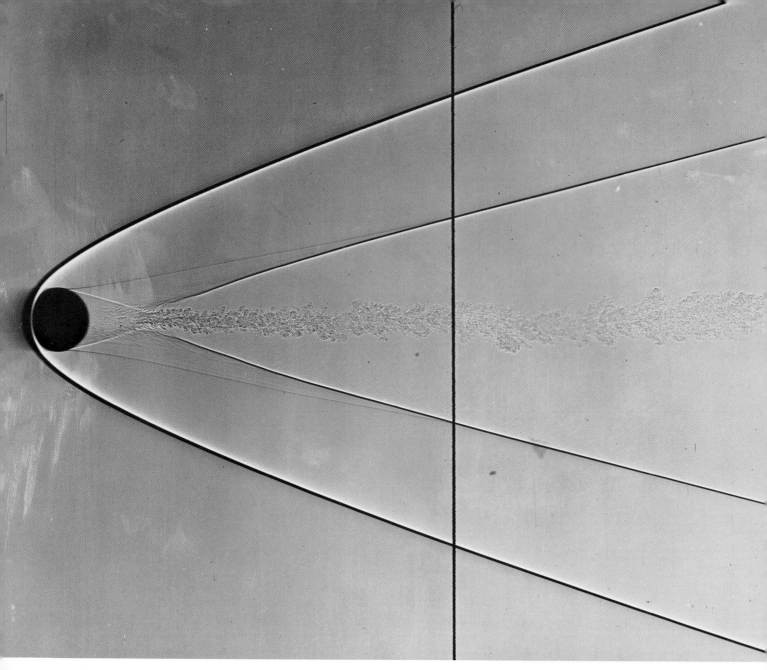

269. Sphere at M=4.01. This shadowgraph of a ½-inch sphere in free flight through atmospheric air shows boundary-layer separation just behind the equator, accompanied by a weak shock wave, and the formation of the N-wave that is heard as a double boom far away. The vertical line is a reference cord. *Photograph by A. C. Charters*

270. Sphere at M=7.6. A nylon sphere is flying through atmospheric air. At this high Mach number the bow shock wave is forced close to the front of the body. Mach waves running downstream from the surface indicate the end of the subsonic region. *U.S. Navy photograph from Naval Surface Weapons Center, Silver Spring, Maryland*

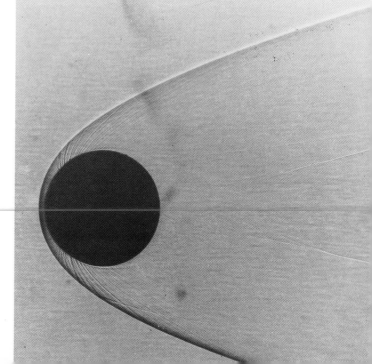

271. Sphere flying over a perforated plate. A shadowgraph shows a $\frac{9}{16}$-inch sphere shot past a plate with a line of $\frac{1}{16}$-inch holes spaced ¼ inch apart. The pressure of the bow wave produces below the plate the classical diagram of the Mach cone as the envelope of a series of expanding spherical acoustic waves. This was used to measure the Mach number, which is seen to be 3. A tiny vortex ring moving downward is formed at each hole, followed at the right by a secondary ring moving upward. *Photograph from U.S. Army Ballistic Research Laboratory*

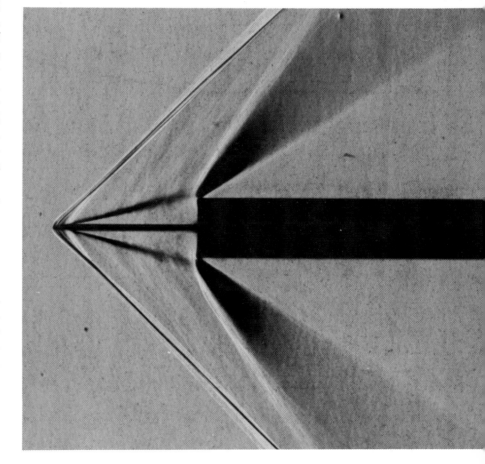

272. Flow past a wedge of dead air. A schlieren photograph shows flow at Mach number 1.96 separating from a thin plate and reattaching downstream at the corners of a thicker plate. The flow pattern resembles that for a solid wedge-shaped leading edge, with a straight bow shock wave followed by a Prandtl-Meyer expansion at the corner. *Photograph by W. A. Mair, courtesy of N. Johannesen*

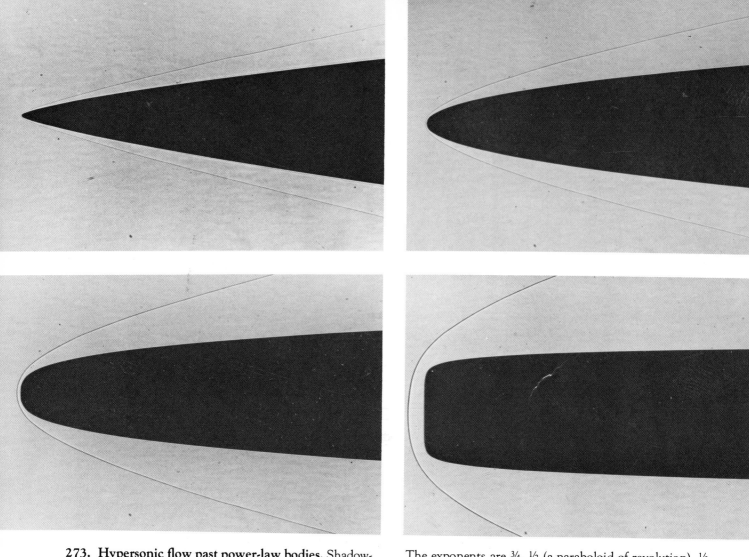

273. Hypersonic flow past power-law bodies. Shadowgraphs show the bow wave in air at M=8.8 for bodies of revolution whose radius varies as a power of axial distance.

The exponents are ¾, ½ (a paraboloid of revolution), ⅓, and ¹⁄₁₀. *Freeman, Cash & Bedder 1964, courtesy of Aerodynamics Division, National Physical Laboratory*

274. Hypersonic flow past a slender cone. A cone of 3° semi-vertex angle is shown by the glow-discharge method in helium at Mach number 41 and Reynolds number 560,000 based on length. In this strong-interaction regime the boundary layer is seen to extend about four-fifths of the distance to the shock wave. *Horstman & Kussoy 1968*

275. Periodic waves from a supersonic jet. An axisymmetric laminar jet of helium from a 0.26-cm converging nozzle into air becomes turbulent within one diameter. It then radiates weak shock waves of frequency 85 kHz, directed primarily along a cone 60° from the axis. The waves are out of phase above and below, the phase shift taking place across the axial plane normal to the page. *Photograph by J. H. Woolley, in Westley, Lindberg, Chan & Lee 1972.*

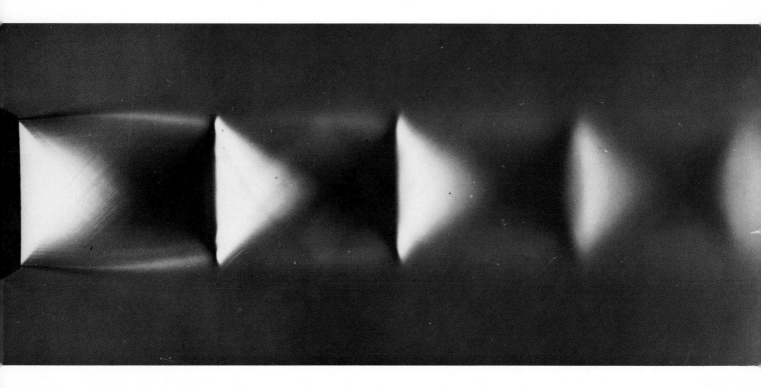

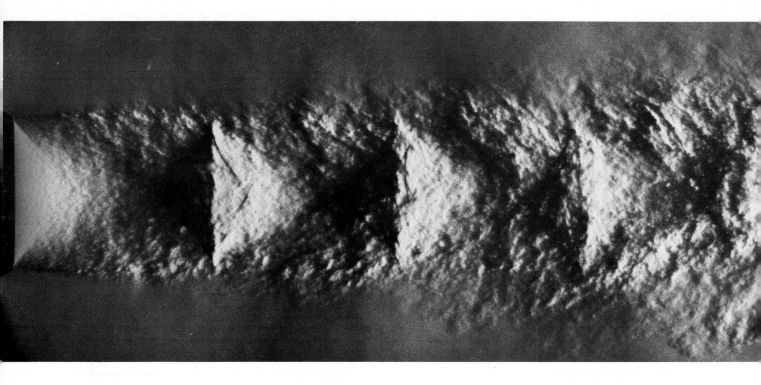

276. Long- and short-duration photographs of a supersonic jet. Dry air flows from a converging conical nozzle with an exit diameter of 1 cm. The ratio of stagnation to atmospheric pressure is 3.13, giving an axisymmetric jet of Mach number 1.4. The upper shadowgraph, with an exposure time of 10^{-2} s, shows the mean flow, with a series of expansion and compression waves. The lower photograph, at 0.5×10^{-6} s, shows the more complicated instantaneous structure. *Photographs by N. Johannesen*

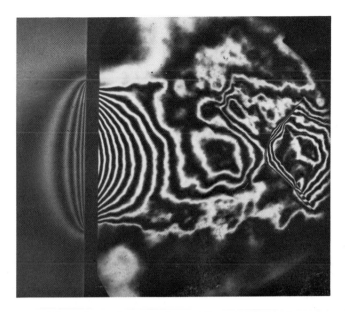
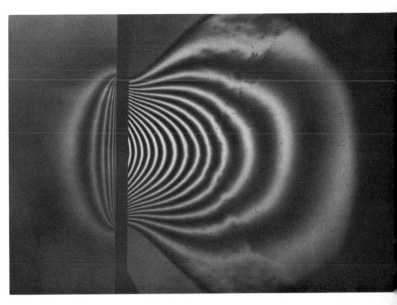

277. Flow through a circular orifice. Air flows through a sharp-edged hole of 5 cm diameter under a pressure ratio of 0.24 in the photograph at the left and 0.0072 at the right. The first interferogram shows a nearly parallel jet downstream of the orifice with strong turbulent mixing and shock waves. The second shows the expansion of a hypersonic jet. *Photographs by W. J. Hiller and G. E. A. Meier*

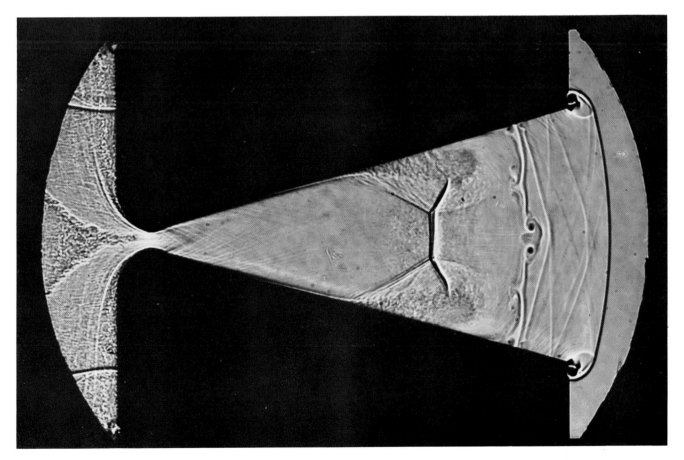

278. Starting process in a nozzle. The incident shock wave, traveling at a shock Mach number of 3, has just passed through a plane nozzle. Behind it are several contact surfaces containing vortices (cf. figure 240). Between them and the nozzle throat is a second shock wave, directed upstream but being swept downstream, and causing the boundary layers to separate. Mach waves from the walls show the supersonic flow established downstream of the throat. *Amann 1971*

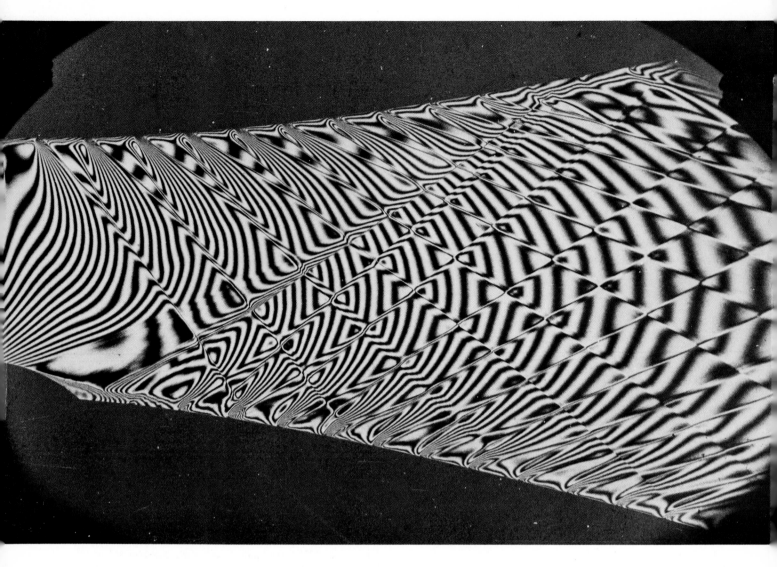

279. Supersonic nozzle with wall disturbances. This interferogram shows the flow through a plane Laval nozzle with a disturbing cylinder on the lower boundary in the throat region, as a model for a disturbed rocket engine. Triangular grooves about 0.3 mm deep have been scratched into the walls to generate weak shock waves (Mach lines). *Hiller & Meier 1975*

References

ACKERET, J., FELDMANN, F. & ROTT, N. 1946 *Mitt. Inst. Aerodyn. Zürich, no. 10*; NACA Tech. Memo. no. 1113, 1947.

ACKERET, J. & ROTT, N. 1949 *Schweizer. Bauz.* 67:40-41, 58-61.

AMANN, H.-O. 1971 *Z. Flugwiss.* 19:393-406.

ASHKENAS, H. I. & BRYSON, A. E. 1951 *J. Aeronaut. Sci.* 18:82-90.

BARDSLEY, O. & MAIR, W. A. 1951 *Philos. Mag.* 42:29-36.

BENJAMIN, T. B. 1967 *Proc. R. Soc. London Ser. A* 299:59-75.

BIPPES, H. 1972 *Sitzungsber. Heidelb. Akad. Wiss. Math. Naturwiss. Kl.*, no. 3, 103-180; NASA TM-75243, 1978.

BLAKE, J. R. & GIBSON, D. C. 1981 *J. Fluid Mech.* 111:123-140.

BLEAKNEY, W., WEIMER, D. K. & FLETCHER, C. H. 1949 *Rev. Sci. Instrum.* 20:807-815.

BRADSHAW, P. 1970 *Experimental Fluid Mechanics.* Oxford: Pergamon Press.

BRADSHAW, P., FERRISS, D.H. & JOHNSON, R. F. 1964 *J. Fluid Mech.* 19:591-624.

BROWN, F. N. M. 1971 *See the Wind Blow.* Dept. Aerospace & Mech. Eng., Univ. Notre Dame.

BROWN, G. L. & ROSHKO, A. 1974 *J. Fluid Mech.* 64:775-816.

BURKHALTER, J. E. & KOSCHMIEDER, E. L. 1974 *Phys. Fluids* 17:1929-1935.

CANTWELL, B. J. 1981 *Ann. Rev. Fluid Mech.* 13:457-515.

CANTWELL, B., COLES, D. & DIMOTAKIS, P. 1978 *J. Fluid Mech.* 87:641-672.

CORRSIN, S. & KARWEIT, M. 1969 *J. Fluid Mech.* 39:87-96.

CORRSIN, S. & KISTLER, A. L. 1954 *NACA Tech. Note no. 3133*; NACA Rept. 1244, 1955.

COUTANCEAU, M. 1968 *J. Méc.* 7:49-67.

COUTANCEAU, M. 1972 *C. R. Acad. Sci. Ser. A* 274:853-856.

COUTANCEAU, M. & BOUARD, R. 1977 *J. Fluid Mech.* 79:231-256.

COUTANCEAU, M. & THIZON, P. 1981 *J. Fluid Mech.* 107:339-373.

CRAPPER, G. D., DOMBROWSKI, N., JEPSON, W. P. & PYOTT, G. A. D. 1973 *J. Fluid Mech.* 57:671-672.

CROW, S. C. 1970 *AIAA J.* 8:2172-2179.

DÉLERY, J., CHATTOT, J. J. & LE BALLEUR, J. C. 1975 *AGARD Conf. Proc.* 168, paper 27.

DÉLERY, J., SURGET, J. & LACHARME, J.-P. 1977 *Rech. Aérosp.* no. 1977-2, 89-101.

DIDDEN, N. 1977 *Mitt. Max-Planck-Inst. Strömungsforsch. Aerodyn. Versuchsanst.* no. 64.

DIDDEN, N. 1979 *Z. angew. Math. Phys.* 30:101-116.

DIMOTAKIS, P. E., LYE, R. C. & PAPANTONIOU, D. Z. 1981 *Proc. XV Int. Symp. Fluid Dyn.*, Jachranka, Poland, to appear.

DYMENT, A., FLODROPS, J. P. & GRYSON, P. 1982 In *Flow Visualization II*, W. Merzkirch, ed., 331-336. Washington: Hemisphere.

DYMENT, A. & GRYSON, P. 1978 *Inst. Méc. Fluides Lille*, no. 78-5.

DYMENT, A., GRYSON, P. & DUCRUET, C. 1980 *Euromech Colloq.* no. 135, Marseilles, France.

ECKERT, E. R. G. & SOEHNGEN, E. E. 1948 *U.S. Air Force Tech. Rept.* 5747.

ECKERT, E. R. G. & SOEHNGEN, E. 1952 *Trans ASME* 74:343-347.

FALCO, R. E. 1977 *Phys. Fluids* 20:S124-S132.

FALCO, R. E. 1980 *AIAA Paper* 80-1356.

FIECHTER, M. 1969 *Jahrb. 1969 DGLR* 77-85.

FREEMAN, N. C., CASH, R. F. & BEDDER, D. 1964 *J. Fluid Mech.* 18:379-384.

FULTZ, D., LONG, R. R., OWENS, G. V., BOHAN, W., KAYLOR, R. & WEIL, J. 1959 *Meteorol. Monogr.* vol. 4, no. 21.

FULTZ, D. & SPENCE, T. W. 1967 In *Atmos. Sci. Paper* no. 122, E. R. Reiter & J. L. Rasmussen, eds., Colo. State Univ.

GEBHART, B. 1971 *Heat Transfer.* New York: McGraw-Hill.

GEBHART, B. & DRING, R. P. 1967 *J. Heat Transfer* 89:274-275.

GEBHART, B., PERA, L. & SCHORR, A. W. 1970 *Int. J. Heat Mass Transfer* 13:161-171.

GLASS, I. I. 1974 *Shock Waves & Man.* Toronto: Univ. Toronto Press.

GRIFFITH, W. C. & BLEAKNEY, W. 1954 *Am. J. Phys.* 22:597-612.

GRIGULL, U. & HAUF, W. 1966 *Proc. 3rd Int. Heat Transfer Conf.* 2:182-195.

HATFIELD, D. W. & EDWARDS, D. K. 1981 *Int. J. Heat Mass Transfer* 24:1019-1024.

HEAD, M. R. 1982 In *Flow Visualization II*, W. Merzkirch, ed., 399-403. Washington : Hemisphere.

HEAD, M. R. & BANDYOPADHYAY, P. 1981 *J. Fluid Mech.* 107:297-338.

HILLER, W. J. & MEIER, G. E. A. 1971 *Bericht 10/1971*, Max-Planck-Inst. Strömungsforsch., Göttingen.

HILLER, W. J. & MEIER, G. E. A. 1975 In E. A. Müller, *50 Jahre Max-Planck-Institut für Strömungsforschung*, Göttingen.

HONJI, H. & TANEDA, S. 1969 *J. Phys. Soc. Jpn.* 27:1668-1677.

HORNBY, R. P. & JOHANNESEN, N. H. 1975 *J. Fluid Mech.* 69:109-128.

HORSTMAN, C. C. & KUSSOY, M. I. 1968 *AIAA J.* 6:2364-2371.

HOYT, J. W. & TAYLOR, J. J. 1977 *J. Fluid Mech.* 83:119-127.

HOYT, J. W. & TAYLOR, J. J. 1982 In *Flow Visualization II*, W. Merzkirch, ed., 683-687. Washington : Hemisphere.

JOHANNESEN, N. H. 1952 *Philos. Mag.* 43:568-580.

JOHANNESEN, N. H. 1958 *Oil* 4:22-26.

KÁRMÁN, T. VON 1947 *J. Aeronaut. Sci.* 14:373-402.

KARWEIT, M. J. & CORRSIN, S. 1975 *Phys. Fluids* 18:111-112.

KLINE, S. J. 1963 *Flow Visualization*, motion picture available from Encyclopedia Brittanica Film Center, Chicago.

KLINE, S. J., REYNOLDS, W. C., SCHRAUB, F. A. & RUNSTADLER, P. W. 1967 *J. Fluid Mech.* 30:741-773.

KOBAYASHI, R., KOHAMA, Y. & TAKAMADATE, Ch. 1980 *Acta Mech.* 35:71-82.

KOSCHMIEDER, E. L. 1966 *Beitr. Phys. Atmos.* 39:208-216.

KOSCHMIEDER, E. L. 1972 *J. Fluid Mech.* 51:637-656.

KOSCHMIEDER, E. L. 1974 *Adv. Chem. Phys.* 26:177-212.

KOSCHMIEDER, E. L. 1979 *J. Fluid Mech.* 93:515-527.

LAUTERBORN, W. 1980 In *Cavitation and Inhomogeneities in Underwater Acoustics*, W. Lauterborn, ed. New York: Springer-Verlag.

MAGARVEY, R. H. & MACLATCHY, C. S. 1964 *Can. J. Phys.* 42:678-683.

MAIR, W. A. 1952 *Philos. Mag.* 43:695-716.

MERZKIRCH, W. 1974 *Flow Visualization*. New York: Academic Press.

MUELLER, T. J., NELSON, R. C., KEGELMAN, J. T. & MORKOVIN, M. V. 1981 *AIAA J.* 19:1607-1608.

NEWMAN, J. N. 1970 *Eighth Symposium on Naval Hydrodynamics*, M. S. Plesset, T. Y. Wu & S. W. Doroff, eds. 519-545. Washington: U.S. Govt. Printing Office.

OERTEL, H., Jr. 1982a In *Flow Visualization II*, W. Merzkirch, ed., 71-76. Washington: Hemisphere.

OERTEL, H., Jr. 1982b In *Thermal Instabilities in Convective Transport and Instability Phenomena*, J. Zierep & H. Oertel, eds. Karlsruhe: Braun.

OERTEL, H., Jr. & KIRCHARTZ, K. R. 1979 *Recent Developments in Theoretical and Experimental Fluid Mechanics*, U. Müller, K. G. Roesner & B. Schmidt, eds., 355-366. Berlin: Springer-Verlag.

OERTEL, H., Sr. 1966 *Stossrohre*. Vienna: Springer-Verlag.

OERTEL, H., Sr. 1975 *Modern Developments in Shock Tube Research*, 488-495. Shock Tube Research Soc., Japan.

PAYARD, M. & COUTANCEAU, M. 1974 *C. R. Acad. Sci. Ser. B* 278:369-372.

PERA, L. & GEBHART, B. 1971 *Int. J. Heat Mass Transfer* 14:975-984.

PERA, L. & GEBHART, B. 1972 *Int. J. Heat Mass Transfer* 15:175-177.

PERA, L. & GEBHART, B. 1975 *J. Fluid Mech.* 68:259-271.

PERRY, A. E. & LIM, T. T. 1978 *J. Fluid Mech.* 88:451-463.

PERRY, A. E., LIM, T. T. & TEH, E. W. 1981 *J. Fluid Mech.* 104:387-405.

PIERCE, D. 1961 *J. Fluid Mech.* 11:460-464.

POLYMEROPOULOS, C. E. & GEBHART, B. 1967 *J. Fluid Mech.* 30:225-239.

PRASSE, H.-G. 1977 *Rept. no. E 10/77*, Ernst-Mach-Institut, Freiburg.

PULLIN, D. I. & PERRY, A. E. 1980 *J. Fluid Mech.* 97:239-255.

ROSHKO, A. 1976 *AIAA J.* 14:1349-1357.

RUTLAND, D. F. & JAMESON, G. J. 1970 *Chem. Eng. Sci.* 25:1301-1317.

RUTLAND, D. F. & JAMESON, G. J. 1971 *J. Fluid Mech.* 46:267-271.

SARPKAYA, T. 1971 *J. Fluid Mech.* 45:545-559.

SAWATZKI, O. & ZIEREP, J. 1970 *Acta Mech.* 9:13-35.

SCHARDIN, H. 1965 *Proc. VII Int. Cong. High Speed Photog.* 113-119. Darmstadt: Verlag O. Helwich.

SCHOOLEY, A. H. 1958 *J. Mar. Res.* 16:100-108.

SEDNEY, R. & KITCHENS, C., JR. 1975 *AGARD Conf. Proc. no. 168*, paper 37.

SHLIEN, D. J. & BOXMAN, R. L. 1981 *Int. J. Heat Mass Transfer* 24:919-931.

SIMPKINS, P. G. 1971 *Nature (London) Phys. Sci.* 233:31-33.

SPARROW, E. M., HUSAR, R. B. & GOLDSTEIN, R. J. 1970 *J. Fluid Mech.* 41:793-800.

STURTEVANT, B. & KULKARNY, V. A. 1976 *J. Fluid Mech.* 73:651-671.

TANEDA, S. 1955 *Rep. Res. Inst. Appl. Mech. Kyushu Univ.* 4:29-40.

TANEDA, S. 1956a *J. Phys. Soc. Jpn.* 11:302-307.

TANEDA, S. 1956b *J. Phys. Soc. Jpn.* 11:1104-1108.

TANEDA, S. 1968 *Rep. Res. Inst. Appl. Mech. Kyushu Univ.* 16:155-163.

TANEDA, S. 1977 *Prog. Aerosp. Sci.* 17:287-348.

TANEDA, S. 1979 *J. Phys. Soc. Jpn.* 46:1935-1942.

TANEDA, S. & HONJI, H. 1971 *J. Phys. Soc. Jpn.* 30:262-272.

TAYLOR, G. I. 1972 In *The NCFMF Book of Film Notes*, 47-54. The MIT Press with Education Development Center, Inc., Newton, Mass.

THORPE, S. A. 1971 *J. Fluid Mech.* 46:299-319.

WALLET, A. & RUELLAN, F. 1950 *Houille Blanche* 5:483-489.

WEGENER, P. P. & PARLANGE, J.-Y. 1973 *Ann. Rev. Fluid Mech.* 5:79-100.

WEGENER, P. P., SUNDELL, R. E. & PARLANGE, J.-Y. 1971 *Z. Flugwiss.* 19:347-352.

WERLÉ, H. 1960a *Rech. Aéronaut.* no. 74:23-30.

WERLÉ, H. 1960b *Rech. Aéronaut.* no. 79:9-26.

WERLÉ, H. 1962 *Rech. Aéronaut.* no. 90:3-14.

WERLÉ, H. 1963 *Houille Blanche* 18:587-595.

WERLÉ, H. 1973 *Ann. Rev. Fluid Mech.* 5:361-382.

WERLÉ, H. 1974 *Le Tunnel Hydrodynamique au Service de la Recherche Aérospatiale*, Publ. no. 156, ONERA, France.

WERLÉ, H. 1980 *Rech. Aérosp.* 1980-5, 35-49.

WERLÉ, H. & GALLON, M. 1972 *Aéronaut. Astronaut.* no. 34, 21-33.

WERLÉ, H. & GALLON, M. 1973 *Houille Blanche* 28:339-360.

WESTLEY, R., LINDBERG, G. M., CHAN, Y. Y. & LEE, B. H. K. 1972 *DME/NAE Quart. Bull.* 1972(1), National Research Council, Canada.

WIJNGAARDEN, L. VAN & VOSSERS, G. 1978 *J. Fluid Mech.* 87:695-704.

WORTMANN, F. X. 1977 *AGARD Conf. Proc.* no. 224, paper 12.

YAMADA, H. & MATSUI, T. 1978 *Phys. Fluids* 21:292-294.

ZDRAVKOVICH, M. M. 1969 *J. Fluid Mech.* 37:491-496.

174

Index

References are to figure numbers unless otherwise noted.